T0336967

Modeling in Food Microbiology

Modeling and Control of Food Processes Set
coordinated by
Jack Legrand, Gilles Trystram

Modeling in Food Microbiology

From Predictive Microbiology to Exposure Assessment

Edited by

Jeanne-Marie Membré
Vasilis Valdramidis

First published 2016 in Great Britain and the United States by ISTE Press Ltd and Elsevier Ltd

ISTE Press Ltd
27-37 St George's Road
London SW19 4EU
UK

www.iste.co.uk

Elsevier Ltd
The Boulevard, Langford Lane
Kidlington, Oxford, OX5 1GB
UK

www.elsevier.com

British Library Cataloguing-in-Publication Data
A CIP record for this book is available from the British Library
Library of Congress Cataloging in Publication Data
A catalog record for this book is available from the Library of Congress
ISBN 978-1-78548-155-0

Printed and bound in the UK and US

Contents

Introduction

The "traditional" food quality and safety management approach is based on end-product testing. However, end-product testing gives only very limited information on the safety status of a food. The food safety crisis of the 1990s (e.g. *Listeria*, *Salmonella*, *Escherichia coli*, *Campylobacter*, dioxins, antibiotics, acrylamide, BSE, etc.) revealed a failure in this traditional approach. If a hazardous organism is found, it means something; but its absence in a limited number of samples is no guarantee for the safety of a whole production batch. End-product testing is often too little and takes place too late [ZWI 16]. Nowadays, effective food quality and safety management systems should rely on prerequisite programs (e.g. good manufacture practices and good hygiene practices), the Hazard Analysis Critical Control Points (HACCP) plan as well as on quantitative tools, namely predictive microbiology and risk assessment approaches.

The development and application of models in food safety and food spoilage falls within the discipline of predictive (food) microbiology, also called predictive modeling in foods [VAN 00, VAN 03], or quantitative microbial ecology [ROS 99]. Schaffner and Labuza [SCH 97] defined this area as the gathering of the disciplines of food microbiology, engineering and statistics to provide useful predictions about microbial behavior in food systems. It, therefore, involves the accumulation of qualitative and quantitative information on microbial behavior in foods and an increased understanding of the microbial physiology [MCM 02a]. Predictive microbiology has a proactive nature in contrast with the retrospective surveillance resting on the prevention of food-borne disease [MCM 02b].

Introduction written by Jeanne-Marie MEMBRÉ and Vasilis VALDRAMIDIS.

According to McMeekin *et al.* [MCM 08], predictive food microbiology in its "modern" form has evolved over the last 30–35 years.

However, microbiological risk assessment is a structured process for determining the public health risk associated with biological hazards in a food. It includes hazard identification, exposure assessment, hazard characterization and risk characterization. Along with the adoption of microbiological risk assessment principles across the globe, since the 2000s, a food safety management based on the risk has emerged (namely, "risk-based food safety management").

Predictive microbiology and exposure assessment are closely connected. Indeed, the models developed in predictive microbiology aim at the quantification of the effects of intrinsic, extrinsic and/or processing factors on the resulting microbial proliferation in food products or food model systems. This enables us to quantify the level of microorganisms in a consumer's portion, information required for exposure assessment. Consequently, both predictive microbiology and exposure assessment play a crucial role in the quantitative implementation of the food safety management system through HACCP, risk-based food safety management and ISO 22000 (International Organization for Standardization). The regulatory authorities have explicitly endorsed the use of predictive tools, e.g. European Commission Regulation (EC) No 2073/2005 referring to *the use of predictive mathematical modeling established for the food in question, using critical growth or survival factors for the micro-organisms of concern in the product.*

Although the concept of predictive modeling introduced from the 1920s with the Bigelow model, which describes the inactivation kinetics of microorganisms, or alternatively, the heat resistance of microorganisms, it took about 60 years to develop models for microbial growth and death. Since then, predictive microbiology has focused on the development of deterministic primary models to predict microbial behavior in foods as a function of storage time (growth and survival) and treatment time (inactivation) with the aid of secondary models which are describing the effect of intrinsic, extrinsic or processing factors on kinetic variables. Despite the progress in this area, the majority of the models developed in the 1990s were based on deterministic approaches without taking into account the variability of factors affecting microbial responses. However, in the context of a risk analysis framework, currently applied at the EU and international level for food safety management, the importance of variability of biological and natural phenomena is widely recognized [FAO 06, ZWI 15]. Deterministic growth or inactivation models that produce single "best" estimates are generally not optimal to satisfactorily manage safety of foods [AUG 12, COU 10, KOU 07]. If, for example, the after effects of unsatisfactory levels of a surviving pathogen in food after processing are grave, the information about the mean population decrease is unlikely to be a sufficient basis to processing design, or if the limits for growth of a foodborne pathogen are not known at a single-cell level, the product formulation for safer food production would be a difficult issue. Furthermore, the use of a worst-case scenario approach in food processing leads to unrealistic estimations with negative impact on food quality. Hence, predictive modeling and exposure assessment are moving towards more sophisticated modeling techniques called probabilistic or stochastic modeling. Stochastic models consider the variation of various factors affecting microbial behavior by using probability distributions of the input variables, and provide the outputs (predictions) expressed as distributions as well [KOY 10, NIC 96, POS 03]. Probabilistic/stochastic modeling is being used with increasing frequency in the area of food safety, and has extensively been utilized in quantitative microbial risk assessment. It should therefore be deployed soon at the operational level, i.e. implemented as a tool in quality and safety management systems.

This book will provide an overview of the major literature in the area of predictive microbiology and probabilistic/stochastic approaches. It will then cover issues related to modeling applications for both microbial spoilage and safety. Food spoilage will be presented through applications of best-before-date determination, commercial sterility, while food safety will be presented through applications of risk-based safety management. Special attention will be paid to probabilistic/stochastic approaches, including model/data uncertainty and biological variability.

Bibliography

[AUG 12] AUGUSTIN J.-C., CZARNECKA-KWASIBORSKI A., "Single-cell growth probability of Listeria monocytogenes at suboptimal temperature, pH, and water activity", *Frontiers in Microbiology*, vol. 3, no. 157, 2012.

[COU 10] COUVERT O., PINON A., BERGIS H. *et al.*, "Validation of a stochastic modelling approach for Listeria monocytogenes growth inrefrigerated foods", *International Journal of Food Microbiology*, vol. 144, pp. 236–242, 2010.

[FAO 04] FAO/WHO (FOOD AND AGRICULTURE ORGANIZATION OF THE UNITED NATIONS/WORLD HEALTH ORGANIZATION), Risk assessment of *Listeria monocytogenes* in ready-to-eat foods, MRA, series 4 & 5, WHO, available at: http://www.who.int/foodsafety/publications/micro/mra_listeria/en/print.html. Accessed 26.05.14, 2004.

[KOU 07] KOUTSOUMANIS K., ANGELIDIS A.S., "Probabilistic modeling approach for evaluating the compliance of ready -to- eat foods with new European Union safety criteria for *Listeria monocytogenes*", *Applied Environmental Microbiology*, vol. 73, pp. 4996–5004, 2007.

[KOU 10] KOUTSOUMANIS K., PAVLIS A., NYCHAS G.-J.E. *et al.*, "Probabilistic model for *Listeria monocytogenes* growth during distribution, retail storage, and domestic storage of pasteurized milk", *Applied and Environmental Microbiology*, vol. 76, no. 7, pp. 2181–2191, 2010.

[MCM 02a] MCMEEKIN T.A., OLLEY D.A., RATKWOSKY D.A.*et al.*, "Predictive microbiology: towards the interface and beyond", *International Journal of Food Microbiology*, vol. 73, pp. 395–407, 2002.

[MCM 02b] MCMEEKIN T.A., ROSS T., "Predictive microbiology: providing a knowledge-based framework for change management", *International Journal of Food Microbiology*, vol. 78, pp. 133–153, 2002.

[MCM 08] MCMEEKIN T.A., BOWMAN J., MCQUESTIN O. *et al.*, "The future of predictive microbiology: strategic research, innovative applications and great expectations", *International Journal of Food Microbiology*, vol. 128, pp. 2–9, 2008.

[NIC 96] NICOLAÏ B.M., VAN IMPE J.F., "Predictive food microbiology: a probabilistic approach", *Mathematics and Computers in Simulation*, vol. 42, pp. 287–292, 1996.

[POS 03] POSCHET F., GEERAERD A.H., SCHEERLINCK N. *et al.*, "Monte Carlo analysis as a tool to incorporate variation on experimental data in predictive microbiology", *Food Microbiology*, vol. 20, pp. 285–295, 2003.

[ROS 99] ROSS T., Predictive microbiology for the meat industry, Meat and Livestock Australia, North Sydney, 1999.

[SCH 97] SCHAFFNER, LABUZA, "Predictive microbiology: where are we, and where are we going?", *Food Technology*, vol. 51, no. 4, pp. 95–99, 1997.

[VAN 00] VAN IMPE J., BERNAERTS K. (eds), "Predictive modelling in foods", *Conference Proceedings*, KULeuven/BioTeC, Belgium, 2000.

[VAN 03] VAN IMPE J., GEERAERD A., LEGUERINEL I. *et al.* (eds), "Predictive modelling in foods", *Conference Proceedings*, KULeuven/BioTeC, Belgium, ISBN: 90-5682-400-7, 2003.

[ZWI 15] ZWIETERING M.H., "Risk assessment and risk management for safe foods: assessment needs inclusion of variability and uncertainty, management needs discrete decisions", *International Journal of Food Microbiology*, available at: http://dx.doi.org/10.1016/j.ijfoodmicro.2015.03.032, 2015.

[ZWI 16] ZWIETERING M., CHIODINI A., JACXSENS L. *et al.*, "Relevance of microbial finished product testing in food safety management", *Food Control*, vol. 60, pp. 31–43, 2016.

1

Predictive Microbiology

1.1. Introduction

Overall, the models developed in predictive microbiology aim at the quantification of the effects of intrinsic, extrinsic and/or processing factors on the resulting microbial proliferation in food products or food model systems, for example, a buffer system (e.g. [WHI 95]). These models rely on the possibility to interpolate the resulting microbial proliferation for combinations that are not only originally examined, but also included in the range of the experiment design. As such, predictive microbiology can be considered as a powerful tool to investigate and summarize succinctly the effect of varying conditions (within food formulation and processing) on the microbial ecology. A historical perspective of modeling developments in the area of predictive microbiology was presented by Mafart in 2005 [MAF 05]. According to that, the first developments date back to 1920s when heat resistance of microorganisms was described by either the Arrhenius equation [ARR 89] or the Bigelow model [BIG 21]. Nevertheless, the principles and goals of the discipline appeared much later, at the beginning of the 1990s, followed by the development and description of microbial models and the generation of relevant databases and other software tools.

An outline of the most recent developments and applications of predictive microbiology as well as some representative published examples is given as follows:

– studies on the assessment of shelf-life (e.g. [XIA 14, HUC 13]);

– applications for process design and optimization (e.g. [VAL 07, KAT 10]);

Chapter written by Vasilis VALDRAMIDIS.

– development of exposure assessment applications (e.g. [TEN 15, MEM 09, PUJ 15]);

– integration of modeling approaches in systems biology (e.g. [BRU 08, VAN 13]).

In the following sections of the chapter, the main principles of predictive microbiology are discussed by providing an overview and interpretation of terms used in the area as well as a classification and a description of the procedure to develop models. Finally, examples of software developments and further literature studies are provided.

1.2. Terminology

A selection of the main terminology in the discipline of predictive microbiology is provided:

– variables: divided in dependent and independent variables. Dependent variables describe a certain response (e.g. microbial population) in relation to some independent variables (e.g. temperature and pH). The independent variables could be also named factors;

– regression analysis: statistical process for estimating the relationship between independent variables and dependent variable, and estimate the model parameters. In predictive microbiology this could involve food extrinsic (e.g. temperature) and intrinsic (e.g. pH and a_w) variables and their relationship;

– prediction: the use of models to forecast responses;

– parameters: set of measurable values which are estimated when the model is solved; they provide the final modeling structure;

– simulation: the use of models to describe responses for a set of predefined variables/factors over time by the use of nominal values. Simulations can be used to perform predictions;

– validation: the process of deciding whether the numerical results quantifying hypothesized relationships between variables, obtained from regression analysis, are acceptable as descriptions of the data. It is an imperative step before using a new modeling structure for other microbiological applications (e.g. performing microbial predictions in food).

The current terminology will help the reader to understand the model development steps and also topics related to the type of models and applications that are available in the literature.

1.3. Classification

Numerous modeling approaches have been published the past years in the literature. These models aim at addressing different issues or interpreting different microbiological phenomena. It is therefore essential for the reader that he/she is first introduced in the area by reviewing how models can be classified based on a number of criteria. These read as follows:

– *Structural characteristics*: based on these characteristics, the models are described as (1) white box or mechanistic (physical) models which are constructed based on theory or underlying mechanisms and are amenable to refinement as knowledge of the system increases [MCM 02], (2) black box or data-driven models are purely based on experimental data and (3) gray box or hybrid models which lay on the interface of white box and black box models, i.e. combining information from both theory and data and having partly interpretable parameters.

– *Kinetic responses*: three model classes in relation to the kinetic responses were defined by Whiting and Buchanan [WIT 93]: primary, secondary and tertiary models. Primary models describe the response of the microbial load as a function of time (growth, inactivation and survival). Secondary models describe the response of primary model parameters to changes in one or more intrinsic, extrinsic and/or processing conditions. Tertiary models are a result of the combination of primary and secondary models together with experimental data, into user-friendly computer software.

– *Categorical responses*: categorical modeling approaches are developed, for instance, to characterize the growth/no growth boundary, and/or to quantify the chance of microbial survival, recovery, or spoilage after certain processing treatments. In between growth and inactivation inducing environmental conditions, the growth/no growth (G/NG) boundary is situated. Generally, it is assumed that this boundary is a transition zone where the growth probability increases from 0% to 100% when going from detrimental to more favorable environmental conditions. These types of model have been developed for different types of assessments including: (1) growth/no growth of pathogenic microorganisms (e.g. [RAT 95, KOU 04, GYS 07]) and of spoilage microorganisms [LOP 00, LOP 07], (2) survival/death of pathogenic microorganism [KOS 07b] and spoilage microorganisms, (3) recovery/no recovery of pathogenic microorganisms (e.g. [KOS 07a, KOS 08]) and finally (4) spoilage/no spoilage [VAL 09].

The development of all the above type of models requires the application of rigorous procedures, which are described in the following sections.

Moreover, the interested users can also refer to a number of books that have been published over the past few years and focus on predictive microbiology and other relevant applications (Table 1.1).

Title: Predictive Modeling and Risk Assessment Author: Rui Costa Publisher: Springer Science & Business Media, 2008 ISBN 0387687769, 9780387687766 Length 170 pages
Title: Advanced Quantitative Microbiology for Foods and Biosystems: Models for Predicting Growth and Inactivation Author: Micha Peleg Publisher: CRC Press, 2006 ISBN: 1420005375, 9781420005370 Length 456 pages
Title: Modeling Microorganisms in Food Editors: S Brul, S Van Gerwen, M Zwietering Publisher: Elsevier, 2007 ISBN: 1845692942, 9781845692940 Length 320 pages
Title: Modeling Microbial Responses in Food Editors: Robin C. McKellar, Xuewen Lu Publisher: CRC Press, 2004 ISBN: 0203503945, 9780203503942 Length 360 pages
Title: Predictive Microbiology: Theory and Application Author: McMeekin, T. A., Olley, J. N., Ross, T. and Ratkowsky, D. A. (1993). Publisher: John Wiley & Sons, New York

Table 1.1. *List of recent books presenting predictive microbiology and its applications in food*

1.4. The modeling cycle

One of the fundamental elements of predictive microbiology and the development of sound modeling approaches is the so-called modeling cycle.

A step-by-step modeling cycle through which predictive microbial models can be rigorously developed is presented hereunder as based on suggested elements in the literature [MCM 93, WAL 97, VAN 98, VAN 99, LJU 99, VAN 01]. The phases of this cycle are illustrated in Figure 1.1 and are described in detail in the following sections.

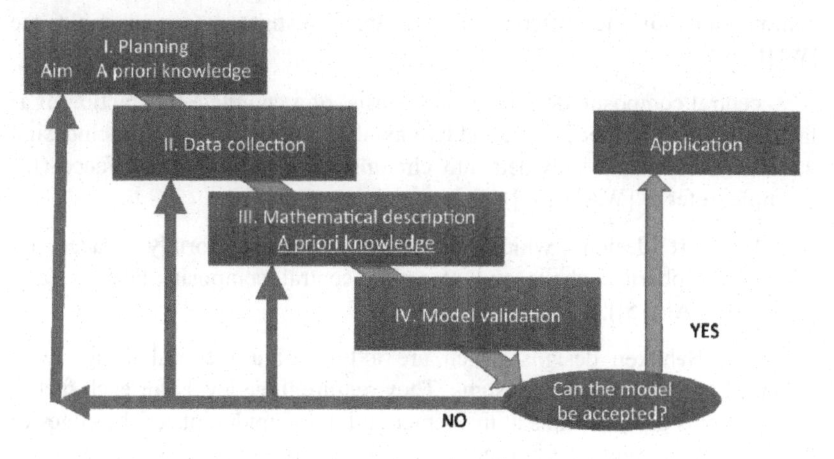

Figure 1.1. *The modeling cycle*

1.4.1. *Planning*

The importance of this first phase is condensed in the dictum of Draper and Smith [DRA 81]: "The specific statement of the problem is the most important phase of any problem solving procedure". Planning involves the gathering of existing microbiological knowledge that may explain and/or simplify observations of the microbial behavior. Moreover, specifying the aim of the model under development (e.g. direct application for determining microbial thermal lethality of specific food products [MUR 03], or a more fundamental study of a microbial inactivation phenomenon e.g. [JUN 03]) largely contributes to a better design in the following steps.

1.4.2. *Data collection*

Informative experimental data and *a priori* microbial knowledge are the means for the development of useful predictive models. An example of a protocol for data generation for predictive microbiology recommends that, for a particular combination of conditions, a minimum of 10 data points should be collected, and the majority of them should be positioned at areas of inflection where the rate of change of the microbial kinetics is maximal [WAL 93]. In cases of studies where more environmental factors are involved multi-factor designs are developed, such as:

– full or fractional factorial experiment design in which all the combinations of the different factors are investigated (for example see [WHI 96]);

– central composite designs which consist of a complete (or fraction of a) factorial design, center points, and two axial points on the axis of each design variable. They can be divided into circumscribed, inscribed and faced (for example, refer to [WAN 13a]);

– Doelhert designs which consist of points uniformly spaced on concentric spherical shells and resemble central composite designs (e.g. [SAU 01, VAL 15]);

– Box–Behnken designs which are a type of a factorial design with balanced incomplete block designs. They require three levels for each factor, only contains combinations at the center and at the midpoints of the edges of the design space and does not include corner points (e.g. [WAN 13b]);

– Latin-square design which is an arrangement of k letters in a k-by-k array so that each letter appears exactly once in each row and exactly once in each column (e.g., [MER 12, DAG 14]);

– optimal experiment design for parameter estimation (OED/PE) which is a novel methodology for the collection of more informative data based on dynamic input profiles (for instance, temperature change(s) during an experiment). (e.g. [WAL 97, BER 02]).

1.4.3. *Mathematical description*

Following the data collection step is the stage of the mathematical modeling where the information collected from the data is interpreted in order to formulate the model design requirements. Considerations that should be taken into account for model development include the following: accuracy of fit, ability to predict untested combinations of factors, incorporation of all relevant factors, possessing the minimum number of parameters, low correlation between parameters, use of parameters with biological meaning and realistic values, reparameterization for improving statistical properties [MCM 93, ROS 95]. In most cases a suitable model has to be selected out of a prespecified set of candidate models, especially when nonlinearity needs to be described. This is due to the lack of generally applicable structure characterization techniques for nonlinear systems [VAN 01]. The careful examination of *a priori* microbiological knowledge may aid in choosing an appropriate mathematical expression. The preselection of a mathematical description could be implemented by first performing a thorough literature review. The users could refer either to papers that review already developed modeling structures (e.g. [GER 05] presents a number of different microbial inactivation kinetic models) or by referring to widely used models (e.g. Baranyi model [BAR 94] for describing the microbial growth).

As mentioned above, one of the main considerations for the mathematical selection is the accuracy of the fit. Therefore, a number of different indices exist to assess this. On the one hand, in the case of kinetic models, these could include, for example, the root mean sum of squared errors, RMSE ([RAT 03] for the appropriateness of this index). On the other hand, for categorical models some of the most common indices are the Akaike Information Criterion, the Schwartz Criterion and the Hosmer–Lemeshow statistic [GYS 07].

1.4.4. *Model validation*

The last (but not least) step of a modeling cycle is the model validation. The application of a mathematical model in more complex systems (different from the laboratory conditions used to develop the model structure and identify its parameters) entails an increase in the error of the predictions [PIN 99].

Three levels of errors can be considered as a result of the different environmental conditions or of the microbial diversity [PIN 99]:

– a primary error due to the difference between the predicted microbial responses and the microbial responses given under similar laboratory conditions (e.g. experiment in a broth system);

– an intermediate error due to difference between predicted microbial responses and microbial responses generated in artificially contaminated foods (e.g. [WIL 02, TSI 03]). According to Miconnet *et al.* [MIC 05], this error can also be due to competition of apathogenic microorganism with a naturally occurring microflora in artificially contaminated foods;

– an overall error due to difference between predicted microbial responses and microbial responses by naturally contaminated food products. This is why generation of an appropriate number of inactivation experiments for determining the model parameters is essential and has to be followed by testing the accuracy of the model with new data using combinations of the examined environmental factors. The step of validation works as an auxiliary methodology to evaluate the goodness of fit, to decide on re-identifying model structure/parameters for improving model accuracy, and consider the necessity of generating additional data.

Models cannot be used with confidence to make predictions in foods until this validation step is successfully passed [WHI 95]. Realistic predictions without being excessively “fail-safe” [PIN 99] are an aim, which cannot be achieved without carefully implementing this step of the modeling cycle. If the model is suitable, it can be used for several applications for which it was developed and validated, otherwise the modeling cycle has to be repeated by re-evaluating decisions made in the different phases. The flexibility and versatility of the developed mathematical models, and precision of the parameter estimates are the appropriate features for accomplishing this final step.

The accuracy and bias factors are considered in order to evaluate the performance capability of the developed (secondary) models. On the one hand, the accuracy factor indicates the spread of results around the simulation or, in other words, how close, on average, simulations are to observations (values close to 1 are indicative of small deviations). On the other hand, the bias factor evaluates whether the observed values lie above the simulation line (i.e. $Bf < 1$, under prediction of microbial load or fail-dangerous) or below the simulation line (i.e. $Bf > 1$, over prediction of microbial load or fail-safe) [ROS 95].

1.5. Software

A number of different software program have been developed which can be used to perform regression or simulations studies and help the user to select appropriate modeling structures and apply the previously described modeling cycle (Figure 1.1). Most of these models are freeware and allow the user to identify the modeling parameters of previously identified modeling structures (i.e. type of solvers for regression) or perform simulation/predictions for given conditions (i.e. type of solvers for simulation).

	Software	Type of solvers	Type of microbial curves	References
1	DMFit	Regression	Growth/ inactivation	[BAR 94]
2	SSSP	Simulation	Growth	[DAL 02]
3	OptiPa	Regression/ Simulation	Defined by the user	[HER 07]
4	GInaFiT	Regression	Inactivation	[GEE 05]
5	UMass	Simulation	Growth/ inactivation	[PEL 08]
6	MRV	Simulation	Growth	[KOS 09]
7	Korean shelf-life decision software	Simulation	Growth	[SEO 09]
8	microHibro	Simulation	Growth/ Inactivation	[PER 13]
9	UGPM	Simulation	Growth	[PSO 11]
10	Fishmap	Simulation	Growth	[ALF 13]
11	PMP	Simulation	Growth/ inactivation	http://pmp.arserrc.gov
12	Package nlstools	Regression	Growth/ inactivation	cran.rproject.org/ web/packages/nlstools/ nlstools.pdf
13	IPMP	Regression	Growth/ inactivation	[HUA 14]
14	PMM-LAB	Regression/ Simulation	Growth/ inactivation	https://sourceforge. net/projects/pmmlab/
15	ComBase			[BAR 04]
15a	ComBase predictor	Simulation	Growth/ inactivation	
15b	DMFit	Regression	Growth/ inactivation	
16	Sym'Previus			[LEP 05]
16a	Growth/bacterial survival simulation	Simulation	Growth/ inactivation	
16b	Growth curve fitting	Regression	Growth	

Table 1.2. *Overview of software for performing model parameter identification and simulation studies (table adapted by [DOL 15])*

1.6. Conclusion

It is evident that over the past few years, a lot of progress has been made in the discipline of predictive microbiology. Knowledge originating from applied statistics, engineering and microbiology is currently integrated within this discipline. Researchers are currently using predictive microbiology tools to perform applications related to shelf-life assessment, process design optimization, risk assessment and systems biology applications. Numerous software are also available so that researchers can develop their own modeling applications as discussed in the previous section.

1.7. Bibliography

[ALF 13] ALFARO B., HERNANDEZ I., LE MARC Y. *et al.*, "Modelling the effect of the temperature and carbon dioxide on the growth of spoilage bacteria in packed fish products", *Food Control*, vol. 29, no. 2, pp. 429–437, 2013.

[ARR 89] ARRHENIUS S., "Über die Reaktiongeschwindigkeit bei der Inversion von Rohmzucker durch Saüren", *Z. Phys. Chem.*, vol. 4, pp. 226–248, 1889.

[BAR 94] BARANYI J., ROBERTS T.A., "A dynamic approach to predict bacterial growth in food", *International Journal of Food Microbiology*, vol. 23, pp. 277–294, 1994.

[BER 02] BERNAERTS K., SERVAES R.D., KOOYMAN S. *et al.*, "Optimal temperature input design for estimation of the Square Root model parameters: parameter accuracy and model validity restrictions", *International Journal of Food Microbiology*, vol. 73, pp. 145–157, 2002.

[BRU 08] BRUL S, MENSONIDES F.I.C., HELLINGWERF K.J. *et al.* "Microbial systems biology: new frontiers open to predictive microbiology", *International Journal of Food Microbiology*, vol. 128, no. 1, pp. 16–21, 2008.

[BIG 21] BIGELOW W., "The logarithmic nature of thermal death time curves, *J. Infect. Dis.*, vol. 29, pp. 528–536, 1921.

[DAG 14] DAGNAS S., ONNO B., MEMBRÉ J.-M., "Modeling growth of three bakery product spoilage molds as a function of water activity, temperature and pH", *International Journal of Food Microbiology*, vol. 186, pp. 95–104, 2014.

[DAL 02] DALGAARD P., BUCH P., SILBERG S., "Seafood spoilage predictor–development and distribution of a product specific application software", *International Journal of Food Microbiology*, vol. 73, nos. 2–3, pp. 343–349, 2002.

[DRA 81] DRAPER N.R., SMITH H., *Applied Regression Analysis*, 2nd ed., John Wiley & Sons , New York, 1981.

[DOL 15] DOLAN K., HABTEGEBRIEL H., VALDRAMIDIS V.P *et al.*, "Thermal processing and kinetic modeling of inactivation", in BAKALIS S., KNOERZER K., FRYER P.J., (eds), *Modeling Food Processing Operations*, Woodhead Publishing Ltd., Cambridge, 2015.

[GER 05] GEERAERD A.H., VALDRAMIDIS V.P., VAN IMPE J.F., "GInaFiT, a freeware tool to assess non-log-linear microbial survivor curves", *International Journal of Food Microbiology*, vol. 102, no. 1, pp. 95–105, 2005.

[GYS 07] GYSEMANS K.P.M., BERNAERTS K., VERMEULEN A. *et al.*, "Exploring the performance of logistic regression model types on growth/no growth data of Listeria monocytogenes", *International Journal of Food Microbiology*, vol. 114, no. 3, pp. 316–331, 2007.

[HUA 14] HUANG L.H., "IPMP 2013-A comprehensive data analysis tool for predictive microbiology", *Interntional Journal of Food Microbiology*, vol. 171, pp. 100–107, 2014.

[HER 07] HERTOG M.L.A.T.M., VERLINDEN B.E., LAMMERTYN J. *et al.*, OptiPa, "an essential primer to develop models in the postharvest area", *Computers and Electronics in Agriculture*, vol. 57, no. 1, pp. 99–106, 2007.

[HUC 13] HUCHET V., PAVAN S., LOCHARDET A. *et al.*, "Development and application of a predictive model of Aspergillus candidus growth as a tool to improve shelf life of bakery products", *Food Microbiology*, vol. 36, no. 2, pp. 254–259, 2013.

[JUN 03] JUNEJA V.K., MARKS H.M., "Characterising asymptotic D-values for Salmonella spp. subjected to different heating rates in sous-vide cooked beef", *Innovative Food Science & Emerging Technologies*, vol. 4, no. 4, pp. 395–402, 2003.

[KAT 10] KATSAROS G.I., TSEVDOU M., PANAGIOTOU T. *et al.*, "Kinetic study of high pressure microbial and enzyme inactivation and selection of pasteurisation conditions for valencia orange juice", *Int. J. Food Sci. Technol.*, vol. 45, p. 11191129, 2010.

[KOS 07a] KOSEKI S., MIZUNO Y., YAMAMOTO K., "Predictive modelling of the recovery of Listeria monocytogenes on sliced cooked ham after high pressure processing", *International Journal of Food Microbiology*, no. 119, no. 3, pp. 300–307, 2007.

[KOS 07b] KOSEKI S., YAMAMOTO K., "Modelling the bacterial survival/death interface induced by high pressure processing", *International Journal of Food Microbiology*, vol. 116, no. 1, pp. 136–143, 2007.

[KOS 08] KOSEKI S., MIZUNO Y., YAMAMOTO K., "Use of mild-heat treatment following high-pressure processing to prevent recovery of pressure-injured Listeria monocytogenes in milk", *Food Microbiology*, vol. 25, no. 2, pp. 288–293, 2008.

[KOS 09] KOSEKI S., "Microbial responses viewer (MRV): a new ComBase-derived database of microbial responses to food environments", *International Journal of Food Microbiology*, vol. 134, nos. 1–2, pp. 75–82, 2009.

[KOU 04] KOUTSOUMANIS K., KENDALL P., SOFOS J., "A comparative study on growth limits of Listeria monocytogenes as affected by temperature, pH and aw when grown insuspension or on a solid surface", *Food Microbiology*, vol. 21, pp. 415–422, 2004.

[LEP 05] LEPORQ B., MEMBRÉ J.M., DERVIN C. *et al.*, "The "Sym'Previus" software, a tool to support decisions to the foodstuff safety", *International Journal of Food Microbiology*, vol. 100, nos. 1–3, pp. 231–237, 2005.

[LJU 99] LJUNG L., *System Identification: Theory for the User*, 2nd ed., Prentice Hall, Inc., Upper Saddle River, New Jersey, 1999.

[LÓP 00] LÓPEZ-MALO A., GUERRERO S., ALZAMORA S.M., "Probabilistic modeling of Saccharomyces cerevisiae inhibition under the effects of water activity, pH, and potassium sorbate concentration", *Journal of Food Protection*, vol. 63, no. 1, pp. 91–95, 2000.

[LÓP 07] LÓPEZ F.N.A., QUINTANA M.C.D., FERNÁNDEZ A.G., "Modelling of the growth–no growth interface of Issatchenkia occidentalis, an olive spoiling yeast, as a function of the culture media, NaCl, citric and sorbic acid concentrations: study of its inactivation inthe no growth region", *International Journal of Food Microbiology*, vol. 117, no. 2, pp. 150–159, 2007.

[MAF 05] MAFART P., LEGUÉRINEL I., COUVERT O. *et al.*, "Quantification of spore resistance for assessment and optimization of heating processes: an ever-ending story", *Food Microbiology*, vol. 27, pp. 568–572, 2005.

[MCM 93] MCMEEKIN T.A., OLLEY J.N., ROSS T. *et al.*, *Predictive Microbiology: Theory and Application*, Research studies Press Ltd., John Wiley & Sons, New York, 1993.

[MCM 02] MCMEEKIN T.A., OLLEY D.A., RATKWOSKY D.A. *et al.*, "Predictive microbiology: towards the interface and beyond", *International Journal of Food Microbiology*, vol. 73, pp. 395–407, 2002.

[MEM 09] MEMBRÉ J.-M., WEMMENHOVE E, MCCLURE P., "Exposure assessment model to combine thermal inactivation (log reduction) and thermal injury (heat-treated spore lag time) effects on non-proteolytic *Clostridium botulinum*", in MARTORELL S., SOARES C.G., BARNETT J. (eds), *Safety, Reliability and Risk Analysis: Theory, Methods and Applications*, vol. 1–4, pp. 2295–2303, 2009.

[MER 12] MERTENS L, VAN DERLINDEN E., VAN IMPE, J.F., "Comparing experimental design schemes in predictive food microbiology: optimal parameter estimation of secondary models", *Journal of Food Engineering*, vol. 112, no. 3, pp. 119–133, 2012.

[MIC 05] MICONNET N., GEERAERD A.H., VAN IMPE, J.F. *et al.*, "Reflections on the use of robust and leat-squares non-linear regression to model challenge tests conducted in/on food products", *International Journal of Food Microbiology*, vol. 104, pp. 161–177, 2005.

[MUR 03] MURPHY R.Y., DUNCAN L.K., DRISCOLL K.H. *et al.*, "Determination of thermal lethality of Listeria monocytogenes in fully cooked chicken breast fillets and strips during post cookin-package pasteurization", *Journal of Food Protection*, vol. 66, pp. 578–583, 2003.

[PEL 08] PELEG M., NORMAND M.D., CORRADINI M.G., "Interactive software for estimating the efficacy of non-isothermal heat preservation processes", *International Journal of Food Microbiology*, vol. 126, nos. 1–2, pp. 250–257, 2008.

[PÉR 13] PÉREZ-RODRÍ GUEZ F., VALERO A., "Predictive Microbiology in Foods", *Springer*, New York, 2013.

[PIN 99] PIN C., SUTHERLAND J.P., BARANYI J., "Validating predictive models of food spoilage organisms", *Journal of Applied Microbiology*, vol. 87, pp. 491–499, 1999.

[PSO 11] PSOMAS A.N., NYCHAS G.J., HAROUTOUNIAN S.A. *et al.*, "Development and validation of a tertiary simulation model for predicting the growth of the food microorganisms under dynamic and static temperature conditions", *Computers and Electronics in Agriculture*, vol. 76, no. 1, pp. 119–129, 2011.

[PUJ 15] PUJOL L., ALBERT I., MAGRAS C. *et al.*,"Probabilistic exposure assessment model to estimate aseptic UHT product failure rate", *International Journal of Food Microbiology*, vol. 192, pp. 124–141, 2015.

[RAT 03] RATKOWSKY D., "Model fitting and uncertainty", in MCKELLAR R., LU X. (eds), *Modelling Microbiology Food Responses*, CRC Press, 2003.

[RAT 95] RATKOWSKY D., ROSS T., "Modelling the bacterial growth/no growth interface", *Letters in Applied Microbiology*, vol. 20, pp. 29–33, 1995.

[ROS 95] ROSSO L., LOBRY J., BAJARD S. *et al.*, "Convenient model to describe the combined effects of temperature and pH on microbial growth", *Applied and Environmental Microbiology*, vol. 61, pp. 610–616, 1995.

[ROS 96] ROSS T., "Indices for performance evaluation of predictive models in food microbiology", *Journal of Applied Microbiology*, vol. 81, pp. 501–508, 1996.

[SAU 01] SAUTOUR M., ROUGET A., DANTIGNY P. *et al.*, "Application of Doehlert design to determine the combined effects of temperature, water activity and pH on conidial germination of Penicillium chrysogenum", *Journal of Applied Microbiology*, vol. 91, no. 5, pp. 900–906, 2001.

[SEO 09] SEO I., AN D.S., LEE D.S., "Development of convenient software for online shelf-life decisions for Korean prepared side dishes based on microbial spoilage", *Food Science and Biotechnology*, vol. 18, no. 5, pp. 1243–1252, 2009.

[TEN 15] TENENHAUS-AZIZA F., ELLOUZE M., "Software for predictive microbiology and risk assessment: a description and comparison of tools presented at the ICPMF8 Software Fair", *Food Microbiology*, vol. 45, pp. 290–299, 2015.

[TSI 03] TSIGARIDA E., BOZIARIS I.S., NYCHAS G.-J.E., "Bacterial synergism or antagonism in a gel casette system", *Applied and Environmental Microbiology*, vol. 69, no. 12, pp. 7204–7209, 2003.

[VAL 07] VALDRAMIDIS V.P., GEERAERD A.H., POSCHET F. *et al.*, "Model based process design of the combined highpressure and mild heat treatment ensuring safety and quality of a carrot simulant system", *Journal of Food Engineering*, vol. 78, pp. 1010–1021, 2007.

[VAL 09] VALDRAMIDIS V.P., GRAHAM W.D., BEATTIE A. *et al.*, "Defining the stability interfaces of apple juice: implications on the optimisation and designof High Hydrostatic Pressure treatment", *Innovative Food Science & Emerging Technologies*, vol. 10, pp. 396–404, 2009.

[VAL 15] VALDRAMIDIS V.P., PATTERSON M.F., LINTON M., "Modelling the recovery of *Listeria monocytogenes* in high pressure processed simulated cured meat", *Food Control*, vol. 47, pp. 353–358, 2015.

[VAN 13] VAN IMPE J.F., VERCAMMEN D., VAN DERLINDEN E., "Toward a next generation of predictive models: a systems biology primer", *Food Control*, vol. 29, no. 2), pp. 336–342, 2013.

[VAN 98] VAN IMPE J.F., VERSYCK K.J., GEERAERD A.H., "Mathematical concepts and techniques for validation of predictive models", in NICOLAI B., DE BAERDEMAEKER J. (eds), *Food Quality Modelling*, Office for Official Publications of the European Communities, Luxembourg, pp. 117–122, 1998.

[VAN 99] VAN IMPE J.F., VERSYCK K.J., GEERAERD A.H., "Validation of predictive models: definitions and concepts", *COST-914: Predictive Modelling of microbial Growth and Survival in Foods*, Office for Official Publications of the European Communities, Luxembourg, pp. 31–38, 1999.

[VAN 01] VAN IMPE J.F., BERNAERTS K., Geeraerd A.H. *et al.*, "Modelling and prediction in an uncertain environment", *Food Process Modelling*, Woodhead Publishing Ltd., Cambridge, pp. 156–179, 2001.

[WAL 97] WALTER E., PRONZATO L., *Identification of Parametric Models from Experimental Data*, Springer-Verlag, Masson, 1997.

[WAL 93] WACKER S.J., JONES J.E., "Protocols for data generation for predictive modelling", *Journal of Industrial Microbiology*, vol. 12, pp. 273–276, 1993.

[WAN 13] WANG J.J., ZHANG Z.H., LI J.B. *et al.*, "Modeling Vibrio parahaemolyticus inactivation by acidic electrolyzed water on cooked shrimp using response surface methodology", *Food Control*, vol. 36, no. 1, pp. 273–279, 2013.

[WAN 13] WANG W., LI M, FANG W.H. *et al.*, "A predictive model for assessment of decontamination effects of lactic acid and chitosan used in combination on Vibrio parahaemolyticus in shrimps", *International Journal of Food Microbiology*, vol. 167, no. 2, 124–130, 2013.

[WHI 93] WHITING R.C., BUCHANAN R.L., "A classification of models in predictive microbiology – a reply to K.R. Davey", *Food Microbiology*, vol. 10, pp. 175–177, 1993.

[WHI 95] WHITING R.C., "Microbial modeling in foods", *Critical Reviews in Food Science and Nutrition*, vol. 35, pp. 467–494, 1995.

[WHI 96] WHITING R.C., SACKITEY S., CALDERONE S. *et al.*, "Model for the survival of Staphylococcus aureus in nongrowth environments", *International Journal of Food Microbiology*, vol. 31, pp. 231–243, 1996.

[WIL 02] WILSON P.D.G., BROCKLEHURST T.F., ARINO S.*et al.*, "Modelling microbial growth in structured foods: towards a unified approach", *International Journal of Food Microbiology*, vol. 73, pp. 275–289, 2002.

[XIA 14] XIAO Lu, FAN X., WANG M. *et al.*, "Shelf-life prediction model of fresh-cut broccoli based on predictive microbiology theory", *Journal of Chinese Institute of Food Science and Technology*, vol. 14, no. 9, pp. 141–146, 2014.

2

Quantifying Microbial Propagation

2.1. Introduction

In our everyday lives, we are faced with many random processes, we may get caught in a rain shower while out walking, bump into an old friend when in town and have a random number of cars pass us by. Because of their random nature, these variables are termed *random variables*. Despite their random nature, these variables can still be characterized by using complex probability theory. Random phenomena can be characterized by a number of probability processes and these provide some element of certainty about random processes. This is particularly important when we consider food safety issues, such as the probability of consuming a contaminated food product, or the probability of importing an infected animal. In such instances, we do not know what the outcome will be, but by using knowledge about probability processes we can calculate the probability of a certain outcome/event. Probability processes help us to better understand and characterize random processes, which is essential when trying to make predictions about a future event or trying to make decisions to reduce/increase the probability of an event.

In considering probability processes and their behavior, we need to consider two types of variable:

– discrete variable: this is a variable which has a discrete number of outcomes. For example, on rolling a dice there are only six discrete outcomes (1–6). Similarly, if we are looking at the number of offspring (calves) from a herd of bovines it can be considered a discrete variable with outcomes 0, 1, 2,

---

Chapter written by Enda CUMMINS.

3 and maximum 4. Hence, the variable is discrete (it is not possible to have 1.5 of a calf). If we are looking at the number of beef burgers contaminated with *E. coli* O157:H7, it is a discrete variable. Discrete variables are usually variables where you can count the number of outcomes;

– continuous variable: this is the second type of variable that we need to consider where the outcome can be an infinite number of outcomes within a range. For example, human height or weight can be considered a continuous variable as these variables can take any value within a range. It is impossible to say all humans will be 5 or 6 feet, etc. They can take any value within a specified range. Continuous variables are usually variables that can be measured.

2.2. Probability processes

2.2.1. *Binomial*

The binomial distribution is used in describing the probability of a discrete random variable. The term "binomial" means two names, and it is used to describe situations where there are just two results, i.e. when there are just two outcomes to an experiment. This is probably the simplest situation that we encounter when faced with an experiment, e.g. the team scores a goal or it does not, if a family has three children they are either boys or girls, on a journey the car crashes or it does not, or in food safety issues the animal is either infected with disease X or not, the food product is either contaminated or not. The probability p is usually taken as the probability of an event occurring while $q = 1 - p$ is taken as the reciprocal probability that the event does not take place. The binomial process covers many real-life situations where there are two outcomes, a "success" or a "failure", and is, therefore, very useful in characterizing the probability of successes or failures from a given number of events (n).

The binomial distribution has the following probability mass distribution:

$$P(X = s) = \binom{n}{s} p^{s} (1 - p)^{n - s} \qquad [2.1]$$

where:

n = no of trails;

s = number of successes;

p = probability of a success s.

Important assumptions of the binomial distribution are:

– each trail is independent of the outcome of the previous trial, or any subsequent trials;

– the probability of a success (p) remains constant for each and every trial.

The binomial considers that in many instances we are not concerned about the exact sequence of successes but are interested in the number of successes from n trials. For example, a hotel buys 100 eggs, the prevalence of salmonella in eggs is 0.001; how many of the purchased eggs have salmonella? The binomial random distribution can be used to characterize such experiments. The binomial random variable thus characterizes the number of successes which occur from n events given that the probability of a success is p. We can use the formula in [2.1] to create the probability density distribution, or in Excel we can use the Binomdist function to calculate it also. The Binomdist function returns the binomial distribution probability for a particular x and has the functional form:

BINOMDIST(number_s, trials, probability_s, cumulative)

Number_s = the number of successes.

Trials = the number of independent trials the experiment is executed.

Probability_s = is the probability of success on each trial.

Cumulative allows the option of computing the specific mass probability (insert “FALSE”) or the cumulative probability (insert “TRUE”) for a particulars.

Take a simple example where a fair coin is tossed 20 times: what is the probability of five heads? We can compute the probability of each outcome using [2.1]. The probability of s successes from n events can be computed easily by hand or, alternatively, the Binomdist function can be used within Excel.

For example, the probability of five heads is:

$$P(X=5)=\binom{20}{5}0.5^5(1-0.5)^{20-5} = 0.0148$$

Or alternatively in Excel we can use the formula:

BINOMDIST(5, 20, 0.5, FALSE) = 0.0148

The same can be done for all possible values of x (1,2,.....20) to create the full probability mass distribution (Figure 2.1).

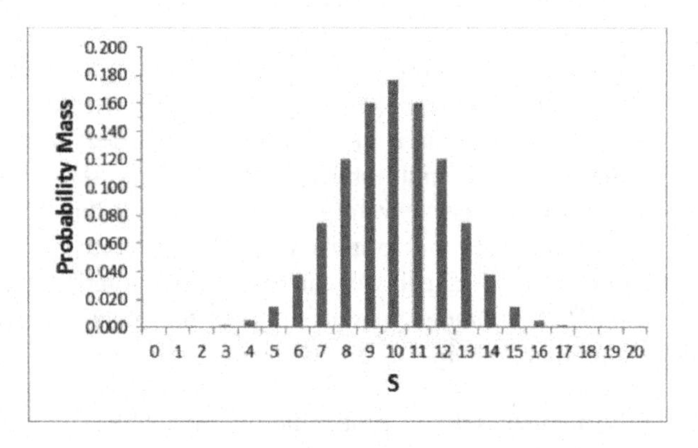

Figure 2.1. *Probability mass distribution (n = 20, p = 0.5), s is the number of successes*

We can also generate the probability mass function using specific Risk Assessment software (e.g. @Risk and Palisade inc.). The general notation of binomial (n, p) is used to signify a binomial process, with specific function available within specialist software to model this. For example, if we were to toss an unbiased coin 20 times how many heads can we expect, the answer is binomial (5, 0.5). The binomial distribution provides the variability distribution for s. The same principles can be applied to food safety issues, as the principles (and assumptions) remain the same.

For example, if 50 cows are imported from country A to country B for breeding purposes and the prevalence of disease X in country A is 0.01. The likely number of imported animals that will have X is given by binomial (50, 0.01) and the probability mass distribution is given in Figure 2.2.

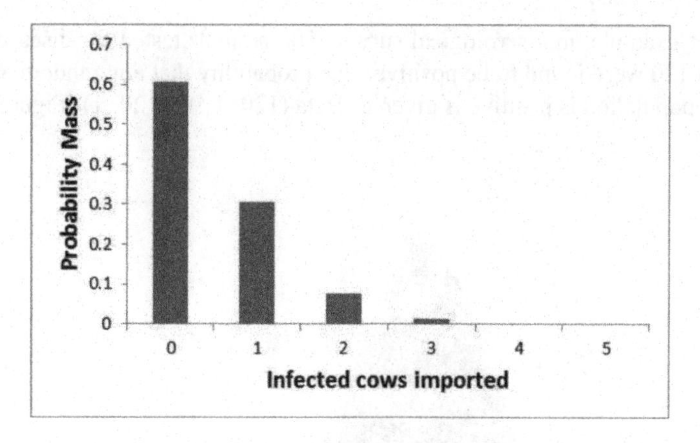

Figure 2.2. *Probability mass distribution for s, the number of infected animals imported (n = 50, p = 0.01)*

2.2.2. *Beta distribution*

The beta distribution is a continuous distribution used to estimate the probability of a success occurring (p) given a number of independent trials (n) and a number of successes (s).

The probability density function of the beta distribution is given by:

$$f(x) = \frac{(a-x)^{\alpha_1-1}(b-x)^{\alpha_2-1}}{Beta(\alpha_1,\alpha_2)(b-a)^{\alpha_1+\alpha_2-1}} \qquad [2.2]$$

where α_1 and α_2 are the shape parameters and a and b are the lower and upper bounds of the distribution and *Beta* ($\alpha 1$, $\alpha 2$) is the beta function [2.3].

The beta function is given by:

$$Beta(\alpha_1,\alpha_2) = \int_0^1 t^{\alpha_1-1}(1-t)^{\alpha_2-1}dt \qquad [2.3]$$

Parameters, $\alpha_1 = s + 1$, $\alpha_2 = n - s + 1$

The beta distribution is often used to model prevalence and provides an uncertainty distribution about p.

For example, in a serological survey, 500 animals tested for disease A in which 120 were found to be positive. The probability that any random animal in the population is positive is given as Beta (120+1,500-120+1) (Figure 2.3).

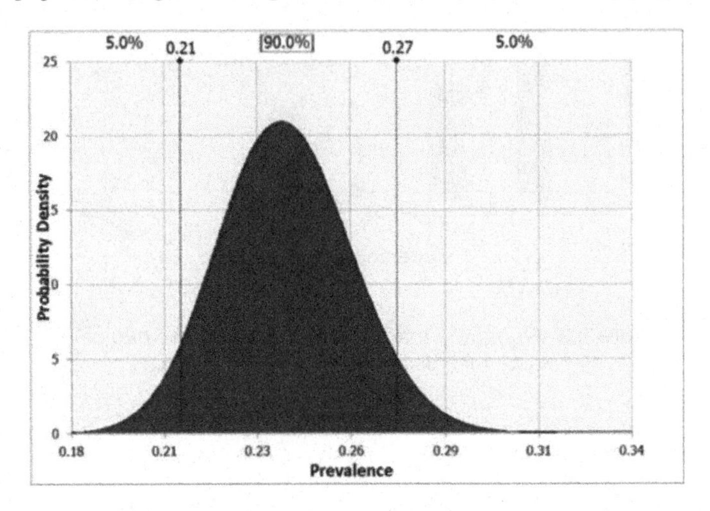

Figure 2.3. *Probability density distribution for the population prevalence of disease A (n =500, s = 120)*

2.2.3. *Negative binomial*

The negative binomial distribution is a discrete probability distribution which estimates the number of failures (*f*) before the sth success in a number of independent trials where each trial has the same probability (*p*) of success. It has the probability mass function:

$$P(X=s)=\binom{s+r-1}{s}p^{s}(1-p)^{r} \qquad [2.4]$$

s = number of success;

r = number of failures before the experiment stops;

p = is the probability of success.

It is a discrete probability distribution which has two input parameters (s and p) and there are two forms of the negative binomial:

1) negative binomial (s, p);

2) negative binomial $(s + 1, p)$.

The first form is taken when we know that we stop sampling after s successes and provides the number of failures before we achieve s successes, the second form is taken when the ordering is not known, but we know that there are just s successes, hence it provides the number of failures before we observe s successes. An example is if you play the lottery scratch card game, where one in five has a cash prize and you want to be assured of five wining cards. There are two scenarios:

1) You buy one ticket and scratch it and continue this cycle until you have the required number of winning tickets (represented by negative binomial (s, p)). You stop buying tickets when you have successfully obtained the five winning tickets, so we know the final ticket is a winning ticket.

2) You go into a shop and bulk buy the tickets (represented by negative binomial $(s+1,p)$) and the order of the winning tickets is not known (i.e. the last ticket may not necessarily be a winning one).

For example, the negative binomial is often used to estimate the number of false negatives. Take the situation where 500 animals were tested for disease X with 95% test sensitivity, five found to be positive. Therefore, using the negative binomial, we can estimate the likely number missed by the test. Number missed = NegBin (5+1,0.95) (Figure 2.4).

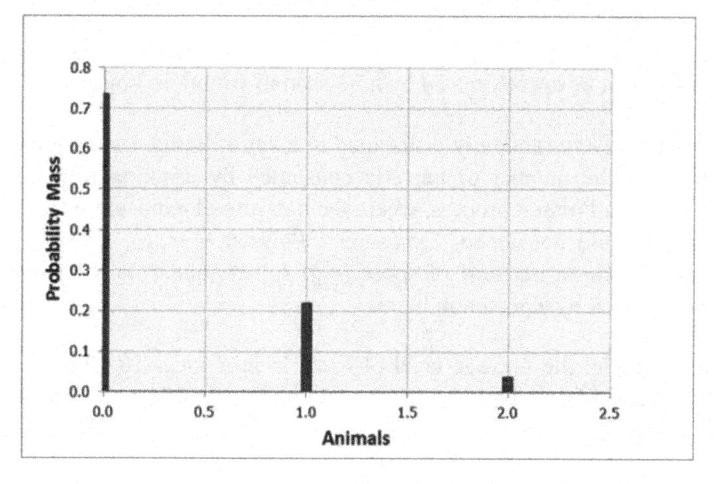

Figure 2.4. *Probability mass distribution for the number of false negatives (s = 5, p = 0.95)*

2.2.4. *Poisson*

The Poisson process is a counting system that counts the number of occurrences of some specific event over a continuum (e.g. time or space). There is a continuous and constant chance for an event to occur in a Poisson process. The Poisson distribution has the probability mass function:

$$P(x=k)=\frac{e^{-\lambda}\lambda^k}{k!} \qquad [2.5]$$

For $k = 0, 1, 2$ .....

λ = mean value of occurrences within an interval (e.g. time or volume).

In a Poisson process, there is a continuous and constant opportunity for an event to occur. The Poisson process could be applied to approximate the number of events that occur in time or space. The Poisson process has a number of assumptions which need to be considered before it use.

The Poisson process assumes:

– the probability of occurrence remains the same for all intervals of equal length;

– the occurrence of an event is independent of whether an event took place before it or indeed takes place after it. It is a memoryless system.

For probability estimates with large n and small p, the binomial distribution can be approximated by a Poisson distribution: Poisson ($n \times p$).

If bacteria were randomly distributed in a vat of liquid, (and not dying or multiplying), the number of bacteria consumed by drinking from that vat would follow a Poisson process, where the measure of exposure would be the amount of liquid consumed, Exposure = Poisson ($c \times L$), where c is the number of bacteria per unit of space (e.g. 1 liter) and L is the amount of liquid consumed by a person in liters.

For example, the average level of bacteria in a vat is 10 CFU/Liter. An individual consumer ingests 2 L of the liquid. The final human exposure can be evaluated as:

Poisson (10×2).

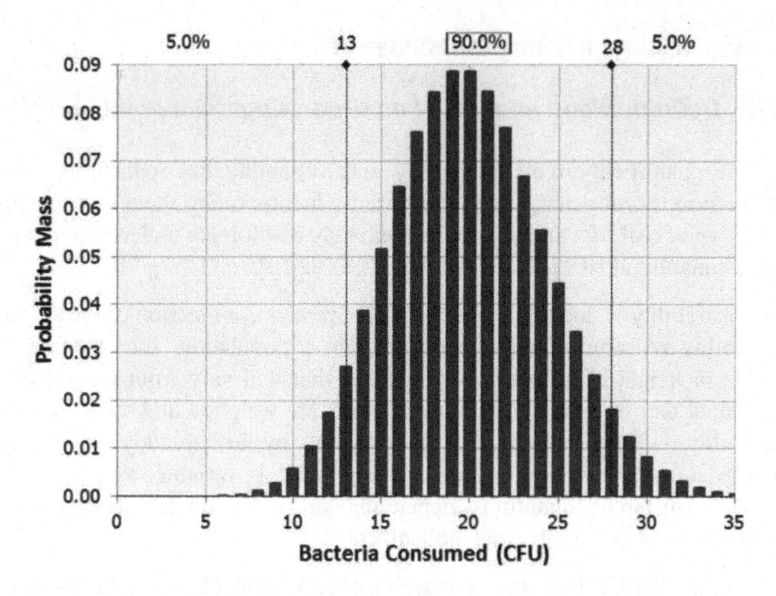

Figure 2.5. *Probability mass distribution for the number of bacteria consumed (λ = 10 × 2)*

We can also make use of the formula in [2.5] to solve problems which relate to the Poisson process. For example, if the average number of bacteria in a vat is 2 CFU/L and if we take 1 liter from the vat what is the probability of having two bacteria in the 1 liter ($k = 2, \lambda = 2 \times 1$):

$$f(2) = \frac{2^2 e^{-2}}{2!} = \frac{.5413}{2} = 0.27067$$

Or in Excel POISSON.DIST(2,2,FALSE) = 0.27067.

Compute the probability of six bacteria in 3 liters ($k = 6, \lambda = 2 \times 3$).

$$f(6) = \frac{6^6 e^{-6}}{6!} = 0.16062$$

Or in Excel POISSON.DIST(6, 6, FALSE) = 0.16062.

2.3. Uncertainty (U) and variability (V)

2.3.1. *Definition and interest of incorporating them in the model*

The quantification of uncertainty and variability has gained increased attention in recent years, simply because by incorporating them both is more reflective of real life, and to neglect them may result in an underestimation or overestimation of risk [GLA 14, CHO 12, POU 10]:

– variability is due to the effects of chance and is a function of the system. Variability represents natural variability in a population, for example the weight or height of a human is something that will vary from individual to individual and no matter how many people are weighed and measured that variability will always exist. It is not reducible by further study. In a similar manner, human consumption of a food product is variable, as people will consume different quantities depending on a multitude of parameters (appetite, level of activity, metabolism, etc.);

– uncertainty refers to our own lack of knowledge about a particular parameter. This is usually due to our inability to accurately measure or capture the required data. For example, in testing for a particular animal disease we use a test which has a less than perfect sensitivity level (e.g. 95% sensitivity). This sensitivity introduces an element of uncertainty in our calculations for the true number of infected animals. We can reduce our uncertainty by perhaps using a more sensitive measurement test which will provide greater confidence about the number of animals infected.

In many instances, the distinction between uncertainty and variability may not be clear cut and may indeed be intertwined. In such cases, total uncertainty is referred to as the uncertainty from both inputs with both variable and uncertain elements:

Total Uncertainty = Variability + Uncertainty.

Hence, some authors treat variability as a certain "type" of uncertainty.

2.4. Modeling propagation using a modular model

When looking at microbial food safety issues, we are interested in transformation of both the prevalence and count of bacteria (Figure 2.6). Microbial growth or cross-contamination is rarely restricted to one individual step in a food chain but is dependent on diligence and safe handling practices across the entire production and processing chain and hence there is a need to look at the entire farm to fork continuum when considering microbial propagation.

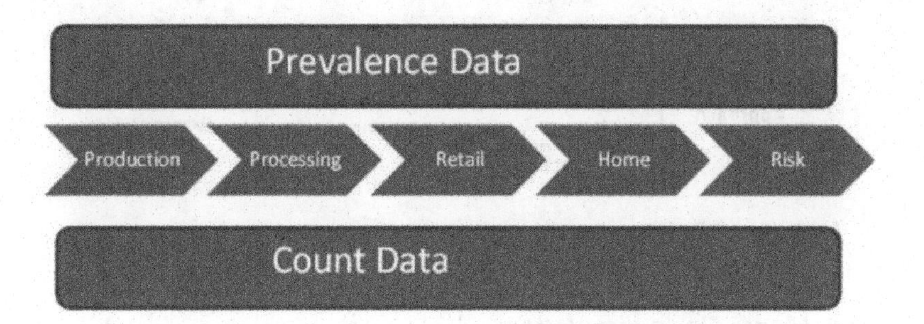

Figure 2.6. *Microbial propagation along the food chain*

The development of mathematical risk assessment models is dependent on adequately capturing both the prevalence and count data of bacterial contamination in a food product and the growth/inactivation of microbes along the chain. Such risk assessment models are dependent on probability processes to capture both prevalence and count data and their behavior along the propagation chain.

2.4.1. *Incorporation of probability distributions into predictive models*

Many models require the monitoring of prevalence and counts of contamination along an entire chain. Several examples exist where a modular approach has been taken to simulate uncertainty and variability from different chains [MOL 15, CUM 08, NAU 07]. In constructing a quantitative model, there are typically two approaches taken (Figure 2.7):

– deterministic: where single value inputs (usually worst-case scenarios) are taken resulting in a single value output;

– probabilistic or stochastic: where the inputs are represented by distributions, therefore, capturing uncertainty and variability in the inputs and resulting in an output distribution. Probabilistic modeling refers to the process of characterizing inputs using probability density distributions and therefore capturing inherent uncertainty and variability.

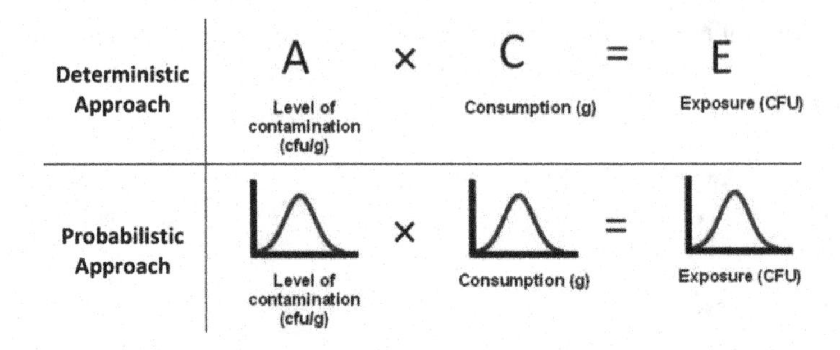

Figure 2.7. *Deterministic and probabilistic modeling approaches*

2.4.2. *Monte Carlo simulation*

The systems we try to model are dynamic and usually have uncertainties and variability associated with many of the inputs. Hence, we need a process which can take account of the variable/uncertain nature of our inputs and reflect this in the output.

Monte Carlo simulation is a process of iteratively randomly selecting a value from a series of probability distributions, carrying out some function with those inputs and detailing the outcome (response variable) of that function. The inputs used are usually detailing elements of uncertainty and variability about the process and therefore Monte Carlo simulation will generate an outcome with a probability distribution which is reflective of the uncertainty and variability in the system. Take, for example, Figure 2.8, there are three inputs represented by distributions, Monte Carlo simulation results in one value being drawn from each input distribution (A1, B1 and C1, respectively), a function is carried out (in this case, A1 is multiplied by B1 and C1 is subtracted) the output of the calculation results in one point on the output distribution (D1). Continuous sampling from the input distributions (termed iterations) results in the output distribution being created. The outcome represents a better representation about the variability/uncertainty of the system and is more reflective of real-life situations. Monte Carlo simulation is particularly useful where direct calculation of a process is difficult, especially where there are several uncertain inputs and when a model is nonlinear. Simulation can typically involve 1,000 s of iterations and with current computing power usually can be carried out very quickly.

A sensitivity analysis can be used to characterize the impact of model inputs.

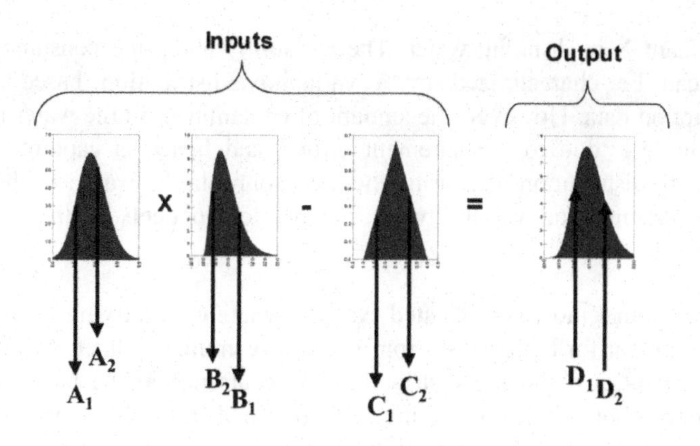

Figure 2.8. *Monte Carlo simulation*

2.5. Separation of uncertainty and variability when building a model

Risk assessment is the qualitative and/or quantitative evaluation of risk while taking account of attendant variability and uncertainty in knowledge and data measurements. Probability density distributions can be used to represent such variability and uncertainty distributions, therefore, allowing adequate capture of the influence of these inputs on the model output [NAU 02].

Given that many systems are highly variable and many inputs are both variable and uncertain, it is important that models try to distinguish between the two processes [GLA 14]. Second-order modeling is the term often given to models where uncertainty and variability have been separated out. By separating the two processes, it may be possible to determine whether uncertainty and variability processes are driving the model. It can provide for greater transparency and identify key driving elements in the model. We know variability is inherent and there is not much we can do about it, but the uncertainty parameters are usually reducible by further study. This has become something of a panacea for risk modelers as separating uncertainty and variability is not always easy.

The use of computer software coupled with probabilistic and Monte Carlo simulation techniques has made the process of separating uncertainty and variability possible, although it still remains quite difficult. Take an example where we are trying to model human exposure in a small town to contaminant X in drinking water. The variability about the consumption of water can be characterized by a variability distribution based around consumption data. However, the amount of contaminant in the water may be uncertain (e.g. due to measurement errors) and hence is captured by an uncertainty distribution measuring the level found in the water. In this way, both uncertainty and variability are inherent components of the exposure model.

Take another more complicated example where we are trying to develop a risk assessment for human risk from bacteria A in minced beef. Again, there are uncertainties about the distribution of occurrence in the beef and also uncertainty about inclusion of minced beef in different food products. In addition, there will be variability regarding possible attenuation and cumulative effects in humans. We can see how the issue increases in complexity and requires a carefully executed systematic approach in dealing with both uncertainty and variability parameters.

Distinction between the two allows more definitive evaluation on whether variability or uncertainty processes are driving the model results and hence allows clearer thinking regarding policy decisions which can be taken to reduce risk [NAU 00]. Where uncertainty processes drive the model more resources can be directed to reduce this uncertainty and improve model estimates, whereas if variability processes are driving the model little can be done to improve the model and outputs merely represent the diversity and range of results which are possible. Of course to separate the two increases the layout complexity of the model and increases the computational complexity also.

2.6. Conclusion

Failure to consider variability and uncertainty can result in erroneous risk assessments and deficiencies in conclusions drawn. A more complete picture can be developed where a second-order modeling approach is adapted and provides greater confidence for policy makers to make informed decisions. The separation of both uncertainty and variability provides greater rigor in terms of data interpretation and data collection requirements. Probability processes can be used to characterize uncertainty and variability in risk assessment models, but knowledge about probability processes and simulation techniques is required.

2.7. Bibliography

[CHO 12] CHOTYAKUL N., PEREZ LAMELA C., TORRES J.A., "Effect of model parameter variability on the uncertainty of refrigerated microbial shelf-life estimates", *Journal of Food Process Engineering*, vol. 35, no. 6, pp. 829–839, 2012.

[CUM 08] CUMMINS E., NALLY P., BUTLER F. *et al.*, "Development and validation of a probabilistic second-order exposure assessment model for Escherichia coli O157:H7 contamination of beef trimmings from Irish meat plants", *Meat Science*, vol. 79, no. 1, pp. 139–154, 2008.

[GLA 14] GLASS K., FORD L., KIRK M.D., "Drivers of uncertainty in estimates of foodborne gastroenteritis incidence", *Foodborne Pathogens and Disease*, vol. 11, no. 12, pp. 938–944, 2014.

[MOL 15] MOLLER C.O.D., NAUTA M.J., SCHAFFNER D.W. *et al.*, "Risk assessment of Salmonella in Danish meatballs produced in the catering sector", *International Journal of Food Microbiology*, vol. 196, pp. 109–125, 2015.

[NAU 00] NAUTA M.J. "Separation of uncertainty and variability in quantitative microbial risk assessment models", *International Journal of Food Microbiology*, vol. 57, no. 1–2, pp. 9–18, 2000.

[NAU 02] NAUTA M.J. "Modeling bacterial growth in quantitative microbial risk assessment, Is it possible?", *International Journal of Food Microbiology*, vol. 38, pp. 45–54, 2002.

[NAU 07] NAUTAM J., "The modular process risk model (MPRM): a structured approach to food chain exposure assessment", *Microbial Risk Analysis in Foods Book Series: Emerging Issues in Food Safety*, 2007.

[POU 10] POUILLOT R., DELIGNETTE-MULLER, MARIE L., "Evaluating variability and uncertainty separately in microbial quantitative risk assessment using two R packages", *International Journal of Food Microbiology*, vol. 142, no. 3, pp. 330–340, 2010.

[TRU 11] TRUDEL D., TLUSTOS C., VON GOETZ N. *et al.*, "Exposure of the Irish population to PBDEs in food: consideration of parameter uncertainty and variability for risk assessment", *Food Additives and Contaminants Part A-Chemistry Analysis Control Exposure & Risk Assessment*, vol. 28, no. 7, pp. 943–955, 2011.

3

Modeling Microbial Responses: Application to Food Spoilage

3.1. Introduction

Roughly, one-third of food produced for human consumption is lost or wasted globally, which amounts to about 1.3 billion tons per year [GUS 11]. For instance, in the European Union, over 100 million tons of food are wasted annually. Food is lost or wasted throughout the supply chain, from initial agricultural production down to final household consumption. One of the causes of food waste is spoilage, leading to food deterioration up to the point in which the food can be no longer consumed. This overall loss of quality could be due to various reasons, which could be sorted into two types: either a change in physico-chemical properties (e.g. textural evolution and production of off-flavor by lipid degradation) or a microbial growth.

Concerning the latter, the range of spoilage microorganisms is wide [BLA 06]. The increasing variety of consumer products together with the need to preserve flavor and texture through minimum processing means that the susceptibility to spoilage has increased, as has the diversity of spoilage species. Bacteria are responsible for some of the most rapid and evident spoilage of proteinaceous foods such as meat and fish products, whereas Gram-positive spore-formers are particularly associated with heat-treated food products [HUI 96]. However, although the growth of yeasts and moulds is generally slower than that of bacteria, the wide variety of ecological niches they can exploit and their tolerance of more extreme conditions than (vegetative) bacteria make them formidable spoilage agents [HUI 96].

Chapter written by Jeanne-Marie MEMBRÉ and Stéphane DAGNAS.

	Inactivation/ reduction	Growth/no-growth interface	Growth (including microbial interaction)
Vegetative bacteria		Seafood and meat product [MEJ 13]	Cantaloupe [FAN 13] Cold-smoke salmon [GIM 04] Ground meat [KOU 09, KOU 06] Minimally processed foods [ZWI 02] Processed seafood and mayonnaise-based seafood salads [MEJ 15] Raw poultry [DOM 07] Ready-to-eat pork product [STA 15] Seafood and meat products [MEJ 13] Shrimp [LER 12]
Spore forming bacteria	Aseptic-UHT products [PUJ 15b] Canned bean green [RIG 14] Canned products [RIG 13] Milk [JAG 03] Tomato pulp [ZIM 13]		Canned bean green [RIG 14]
Yeast and mould	Apple juice [MAN 11, SAN 09] Canned tomato paste [KOT 97]	Sugar-based product [DES 15] Sweet and sugar-based product [MAR 15]	Bakery products [DAG 14, GUY 05, HUC 13, ROS 01] Beverages [BAT 02] Corn [AST 12, SAM 07b] Grape berries [JUD 10] Green table olives [PAN 03] Yogurt [GOU 11]

Table 3.1. *List of recent, but not exhaustive, applications of modeling to food spoilage, sorted by types of spoilage microorganisms and type of models used to describe the microbial behavior*

In terms of modeling, food spoilage has been quantified through inactivation models, growth/no-growth models and growth models. The mathematical structure of the predictive models (Chapter 1) is similar to what has been developed in the microbial modeling area for more than 30 years [BRU 07, DAN 13, MCK 04, SME 14, VAN 98]. Nevertheless, the applications of modeling to food spoilage are less numerous than in safety (Table 3.1), even if in several categories of food products, it is more difficult to control the spoilage microorganism than the pathogenic microorganism. A recent study focused on aseptic ultra-high-temperature (UHT)-type processed food products illustrates clearly this latter point [PUJ 15a]. The authors have shown that *Clostridium botulinum* was eliminated at the heat-treatment step (in their example, based on dairy product line, an F_0 value of 21 min was applied), while the risk of commercial sterility failure due to *Geobacillus stearothermophilus* was not negligible (estimated mean risk value of 600×10^{-8} per product unit, value which could appear as low, but which is not considering the huge market of dairy products treated by aseptic UHT process).

Section 3.3 of this chapter will focus on mould spoilage of bakery products [DAG 14, GUY 05, HUC 13]. In this type of product, in case of air recontamination (between baking and packaging steps), the water activity (a_w) of the product is too low to enable bacterial growth; on the contrary, mould spores can germinate and grow to a visible mycelium leading to product rejection by the consumer. Consequently, microbial modeling has been focused on quantifying the growth response (germination, lag and mycelium growth) with the objective of understanding how the formulation (mostly a_w, as well as the addition of preservatives) and to a lower extent the environmental conditions (storage temperature) could prevent mould growth. In this context, it will be shown here how predictive models and probabilistic approaches can be deployed to aid in formulation design and also in best-before-date (BBD) determination, since with a mycelium proliferation phenomenon the consumer rejection depends on the time of storage.

3.2. Modeling spoilage: application to commercial sterility

3.2.1. *Context: commercial sterility and G. stearothermophilus*

UHT-processed and canned foods are ready-to-eat products which are consumed extensively globally due to a combination of their comparative high quality and long shelf life, with no cold chain or other preservation requirements. Commercial sterility of such food products corresponds to the condition achieved by application of sufficient heat to render the food free from microorganisms capable of growing in the food under ambient storage conditions [COD 93]. The process has traditionally been run using deterministic or empirical process settings, with the aim of achieving a low sterility failure rate, at least lower than the currently tolerable sterility failure rate (i.e. one defect per 10,000 units). There are only few quantitative studies focused on commercial sterility [AND 11, CER 01, MEM 11, PUJ 15a, RIG 14].

G. stearothermophilus (formerly, *Bacillus stearothermophilus*) is a Gram-positive spore forming bacterium. Due to its capacity to survive very high heat treatment [GHA 02, RIG 13, STU 73], *G. stearothermophilus* has often been reported as responsible for ambient stable product spoilage [DEN 81, ITO 81, PRE 10]. More precisely, Andre *et al.* [AND 13] have determined that *Geobacillus* sp. was involved in 35% of non-stability cases of canned food. In industry, before factory release, a commercial sterility test is performed by checking stability of end-product at 30°C (for 10±3 days) and 55°C (for 6±1 days). This latter temperature is set to pick whether thermophilic bacteria, such as *G. stearothermophilus*, are present in the end-product.

3.2.2. *Modeling framework: modular process risk model*

When conducting a quantitative microbial risk assessment (QMRA) with a food safety application in mind, the exposure assessment step objective is to determine the mean and distribution of microorganisms in the food at the moment of consumption. Consequently, in food safety, the concentration of microorganisms is often characterized by analyzing statistically experimental food samples at the time of consumption or to a step before, for instance at the retail point [HAA 14]. For instance, in their study on *Salmonella* in chicken meat preparations mandated by the national authorities, Uyttendaele *et al.* [UYT 09] started their exposure assessment when the food arrives at the retail level. Likewise, the FAO and the WHO have commissioned a study aiming to characterize the risk of *Listeria monocytogenes* in ready-to-eat foods in which the focus of the exposure assessment models was to account for changes in the frequency and extent of contamination in the food between retail marketing and the point of consumption [FAO 04].

On the contrary, for spoilage, the exposure assessment step is more comprehensively studied: the contribution of each process step during the food transformation and/or distribution is quantified with the ultimate goal of identifying and evaluating management options to control or reduce this risk. Note that at least in Europe, using the word "risk" to describe the probability of having a food unfit for human consumption is in agreement with the Food Law of the European Commission. Indeed, the Food law specifies that a food product is unsafe if it is unacceptable for human consumption according to its intended use, for reasons of contamination, whether by extraneous matter or otherwise, or through putrefaction, deterioration or decay [EUR 02].

To quantify accurately the effect of the various process steps on the exposure assessment, Nauta [NAU 01] introduced the "Modular Process Risk Model" concept where the food pathway is divided into several key smaller steps.

In an aseptic-UHT process, the contamination pathways include (1) the introduction of microorganisms with the raw materials, (2) the sterilization step (for the product and packaging), (3) the potential recontamination of the product due to intermediate storage and transport in pipelines (recontamination due to biofilm formation or from air), (4) the partitioning and filling step, with again potentially a recontamination and finally (5) storage of end-product at ambient temperature for several months/years [PUJ 15b].

With a canned vegetable-based food, the contamination pathways include (1) the introduction of microorganisms with the raw materials, (2) the blanching where there is a partial inactivation of spores as well as a potential recontamination by blanching water, (3) the canning and brining step, with also a possibility of recontamination by recovery brine, and finally, (4) sterilization before (5) final storage for several months/years [RIG 14].

The common and key steps which have to be taken into account considering the risk of commercial sterility failure due to *G. stearothermophilus* are (1) the introduction of microorganisms with the raw materials and (2) the sterilization step; the recontamination pathways vary in the two processes: post-thermal treatment in the case of UHT process, prior-thermal treatment in the case of canned product. The mathematical expressions describing these different microbial contamination pathways are presented in Table 3.2.

Contamination pathway	Context/ assumptions	Recommended or commonly applied modeling approaches	Case-studies used here for illustration purpose
Introduction through the raw materials	In the case of spoilage, the initial level of contamination is often high, meaning that the focus is on modeling the concentration; prevalence is assumed to be 100%	Concentration: normal or lognormal distributions	Normal distributions [RIG 14]
Reduction in concentration and prevalence due to sterilization	Log reduction: ratio of bacterial concentration before and after the process step, expressed in decimal logarithm (log cfu/g or log cfu/ml)	Primary model: often a simple first-order kinetic model [BIG 21] is used in industrial applications. See Chapter 1 for alternatives	First-order kinetic model used in both green bean and milk-based product applications [PUJ 15b, RIG 14]
	Number of surviving spores and post-process prevalence	Interpretation of primary model output is probabilistic, using a binomial distribution or its approximation to a Poisson distribution [MEM 07, NAU 01, POU 15]	Poisson distribution was used in the case of a milk-based product [PUJ 15b]
	Decimal reduction time (D). This depends on the temperature applied and also in some cases of the formulation (pH and water activity) or the food matrix itself	Different secondary model structures have been developed to describe D with respect to the process (heat-treatment temperature) and product formulation (pH and water activity (a_w)). See Chapter 1 for more details	Model with only the temperature effect in the case of a milk-based product [PUJ 15b], model with temperature and pH in the case of green bean [RIG 14]
Modeling recontamination (prior-thermal process)	In the canning process, additional contamination, besides the introduction with the raw materials, could occur, before the heat-treatment step, for instance due to brining operation	The golden rule with recontamination is to consider the addition of contaminants in an arithmetic sum. Do not sum microbial count after logarithm transformation	Normal distribution to estimate the microbial concentration in blanching water and brine [RIG 14]

Modeling recontamination (post-thermal process)	Modeling recontamination from air: The recontamination from air is a key issue to control for a UHT-aseptic process because the "sterile" product could be exposed to the air, e.g. in the area of the filling machine	Contamination level on the product [DEN 03]: $L_c = C_{air} \times v_s \times A \times t$ [3.1] where: L_c is the number of microorganisms which contaminate a product unit; C_{air} is the concentration of microorganisms in the air (CFU/m^3); v_s is the settling velocity (m/s); A is the exposed area of a product unit projected on the horizontal plane (m^2); t is the exposed time (s) C_{air} depends on microbial concentration outside the factory, efficiency of filters, frequency of their replacement, positive pressure applied, etc. [DAG 13]	Number of spores contaminated the product at the filling step estimated using [3.1] [PUJ 15b]
	Modeling recontamination from biofilms: In UHT-aseptic process, microorganisms could attach to the surfaces of storage tank, pipes, valves, etc., during a batch, and be released at the next batch (between or after in-place cleaning operations)	The quantity of microorganism in the biofilm is modeled with a mass balance equation [DEN 03]: Biofilm quantity = Adsorption – Release + Growth The adsorption depends on the quantity of microorganisms initially in the product (which settle down to the surface) as well as on the complexity of the process line (e.g. number and cleanability of pipe corners and valves, number of pipe diameter changes, etc.).	A mass balance equation was applied to describe the potential contamination due to biofilm formation; the adsorption model was simple in the first version of the model (transfer rate with a huge uncertainty, see [PUJ 15b]) and refined in the second one to include elements of equipment design [PUJ 15c]

Table 3.2. *Modeling approaches which are, or could be, used to quantify the microbial behavior for the key process steps associated with commercial sterility failure due to G. stearothermophilus*

3.2.3. *Model set up: choices of inputs and simulation procedure*

In a modular process risk model, several types of inputs have to be distinguished:

– the first one is directly link to management options, such as formulation and process settings. Their value can be changed to reduce the risk of commercial sterility failure. For instance, the temperature of the UHT treatment is a thermal process setting while the pH is a formulation setting. When building the model, it makes sense to set them to a single value (for instance, to an average value in a baseline scenario representative of a generic factory line), to be able, later on, to compare what the model output is when the setting is fixed to another value, in a what-if scenario analysis. In other words, the settings are deliberately considered as deterministic inputs;

– the second type corresponds to deterministic inputs which are not settings, because it is impossible to associate any management option with these inputs. For instance, the density of raw materials is a variable required in the model, which cannot be used to control and/or reduce the risk. They are known precisely, the uncertainty is negligible, reason why they are set to a fixed value;

– the third type corresponds to probabilistic inputs. They are set to a range of values with their associated probability of occurrence. Probabilistic inputs could reflect variability, uncertainty or both. The variability captures the biological diversity [THO 02], for example the diversity in *G. stearothermophilus* strains heat resistance (D-value at 121°C, D_{121}). The uncertainty captures the lack of knowledge (lack of data and lack of certainty of subject-matter experts of the domain), for example, the lack of data/knowledge to build precisely the distribution of *G. stearothermophilus* D_{121}. In the model developed for canned green bean [RIG 14], log D_{121} variability was described by a normal distribution N(μ, sd), in which μ was described by a normal distribution, and sd was described by a lognormal distribution, to take the uncertainty dimension into account. In the model developed for UHT-dairy product, the log D_{121} variability was described by a Pert distribution, with the three parameters (5th percentile, most likely value and 95th percentile) described by uniform distributions to take the uncertainty dimension into account.

3.2.4. *Probabilistic toolbox: second-order Monte Carlo analysis, sensitivity and scenario analysis*

The separation of uncertainty and variability has been recommended for more than a decade [NAU 00]. Once separated, the model is run using a second-order Monte Carlo technique, leading to build a confidence interval around a risk estimate for different realizations of variability; that enables us to identify if further investigation on uncertainty is required before making any decision [MOK 05]. A few examples of its application are available in the literature for pathogenic bacteria but only a limited number with spoiling microorganisms. One of the reasons is that second-order Monte Carlo technique is not straightforward and can be time-consuming; it has to be fit-for-purpose and deployed only when risk assessors/risk managers have identified its true benefit [MEM 12].

In the study on UHT-dairy product, the second-order Monte Carlo was implemented in Microsoft Excel using the @RISK 6.3.1 software (Palisade Corporation). First, a sample of 1,000 values from each uncertainty distribution was generated. Next, for each realization of uncertainty, a simulation was run using the "risksimtable" function of @risk. For each simulation of the variability dimension, 10,000 iterations were generated. In the study of canned green beans, the second-order Monte Carlo was implemented in R software (R Development Core Team, 2010), using the mc2d package [POU 10]. A sample of 10,000 values was used for the uncertainty dimension and a simulation of 100,000 iterations was generated for the variability dimension. In both examples, the sample size, in both uncertainty and variability dimension, is limited. This obviously has an effect on the model output accuracy, which is why it is essential to evaluate the model output accuracy before interpreting the results. This could be easily done by running few simulations (with independent seeds) and comparing the results.

Sensitivity analysis is widely used in quantitative risk assessment to (1) identify the most influential variables in a model, (2) provide a better understanding and interpretation of the analysis and (3) identify data gaps and then prioritize future research. Sensitivity analysis might be seen, as well as a prerequisite for model building, for testing the robustness and relevance of the model [SAL 02]. In the case studies detailed here, sensitivity analysis was done. Unsurprisingly, despite a lot of process operation units, and probabilistic inputs, both studies concluded that the thermal treatment had the highest influence on the commercial sterility failure. In other words, to control or even reduce the risk, the heat-treatment setting (time and temperature, or equivalent) was a crucial point.

More interestingly, scenario analysis was also run in both case studies. For canned green beans, the median risk of non-stability due to *G. stearothermophilus* varied from 0.5% [0.1%; 1.2%] in the baseline scenario (F_0 value of 30.2 min) to 2.0% [0.8%; 3.9%] with an F_0 value of 20 min. Note that in both cases, the risk is relatively high and very far from the generally accepted tolerable limit (1 in 10,000 product units).

With the UHT-dairy product, the mean risk of non-complying with commercial sterility due to *G. stearothermophilus* was assessed to 580 $\times 10^{-8}$ [430 $\times$ 10^{-8}; 750 $\times$ 10^{-8}] under the base line scenario (138 °C for 25 s) to 20 $\times$ 10^{-8} [10 $\times$ 10^{-8}; 34 $\times$ 10^{-8}] under a higher setting (138 °C for 31 s). Considering that a risk of 20 $\times$ 10^{-8} was acceptable for *G. stearothermophilus*, various combinations of holding time and temperature were calculated (Figure 3.1). They correspond to combinations of management options to control the risk.

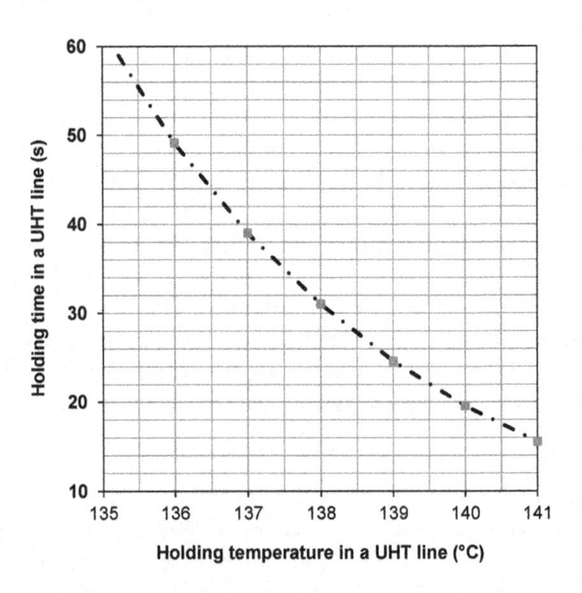

Figure 3.1. *Combinations of holding time and temperature in a UHT line enabling us to control the risk of commercial sterility failure, due to Geobacillus stearothermophilus to an acceptable level. Adapted from [PUJ 15a]*

3.3. Modeling spoilage: application to best-before-date determination

3.3.1. *Context: mould spoilage of food products*

Moulds are able to grow within a wide range of water activity (a_w), pH and temperature by using a large diversity of substrates such as carbohydrates, organic acids, proteins and lipids [HUI 96]. Consequently, moulds are able to grow on acidic products such as fruits and fruits juices

[LAH 05] or intermediate moisture foods such as bread and bakery products [ABE 99], whereas other microorganisms such as bacteria are not.

For manufactured products, food spoilage by fungi results from in-factory fungal contamination followed by mould growth on food products (Figure 3.2), this latter including two interconnected phenomena: spore germination and mycelium growth [DAG 13]. The development of visible mycelia on the surface of a product leads to its rejection by the consumer [GOU 11, HOR 73].

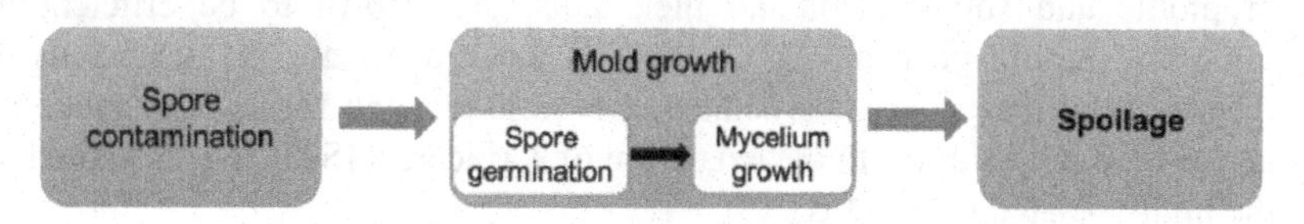

Figure 3.2. *Schematic representation of phenomena involved in food spoilage by moulds*

Mathematically, the spoilage can be expressed as a combination of probabilities: probability of being contaminated and probability of growing (germination and mycelium growth) to a visible mycelium before product consumption. Each probability is described by a statistical model with a response (the probability) and several factors of variation. These factors are introduced in the model as single values when a deterministic approach is preferred or as a probability distribution function when a probabilistic approach is chosen. In the case of mould spoilage assessment, they are variability in some process and environmental factors; for instance, the storage temperature varies with the region, the season and also the consumer's storage habits. Likewise, the mould growth characteristics vary with the mould species or even with the mould strain. Storage temperature and growth characteristics should be then introduced in the assessment model as probability distributions. Overall, this means that food spoilage by fungi is preferably described by a probabilistic framework.

3.3.2. *Factors affecting the mould growth on bakery-type products*

In terms of effect on mould growth, water activity (a_w) is widely recognized to have the greatest effect on spore germination and mycelium growth [PIT 77]. The a_w depression leads to a decrease in the speed of both germination and growth until a minimum is reached, minimum where neither germination nor growth occur. The temperature is recognized as the second most important factor to have an impact on mould growth after a_w. Among moulds, some species are able to grow at moderate and high temperature. For instance, *Aspergillus flavus* and *Aspergillus niger* are able to grow between 8 and 45°C with an optimum near 30°C [PIT 75]. These moulds are involved in spoilage of shelf stable food, especially in hot weather.

In order to prevent food spoilage from microorganisms, the use of preservatives is almost a routine for manufacturers. Among them, acetic, propionic and sorbic acid, and their salts, are known to be efficient in inhibiting mould growth [GUY 02]. The weak acids are pH-dependent as they exist in two states: dissociated and undissociated forms. The mode of action of weak acids (with the exception of sorbic acid [STR 09]) is related to the undissociated state [DAG 15].

The inhibitory effect of ethanol has been reported on both spore germination [DAO 10] and mycelium growth [DAN 05a] in either stopping growth or delaying it by disrupting the cell membrane. In manufacturing food product, ethanol is used directly on the product (e.g. sprayed on the surface) or in the packaging atmosphere (ethanol vapor) [DAO 11]. Finally, the moulds encountered in food spoilage are strict aerobes, i.e. oxygen is necessary for mould growth.

3.3.3. *Primary and secondary models developed (specifically) for mould spoilage*

At an infancy stage, predictive microbiology was mainly focused on bacteria. Later on, predictive microbiology was applied to mould spoilage. Because of (or more exactly due to) this chronological order, some models early developed for bacteria were just slightly adapted to fit mycelium growth curves, or to describe the influence of T, pH and a_w on mould growth rates. Having said that, the cardinal value model developed by Rosso [ROS 01] to quantify the influence of a_w on mould growth was one of the first cardinal value models ever published.

Nevertheless, there are some differences between bacterial growth prediction and mould growth prediction. The most important difference concerns the germination: some specific models have been developed to describe the germination process (Table 3.3).

Mould growth stage	Models	Examples of applications
Germination	Gompertz equation $$P_{(t)} = P_{max}.exp\left(-exp\left[\frac{\mu_G.e(1)}{P_{max}}(\lambda_G - t) + 1\right]\right)$$ [3.3] where t (h) is the time, $P_{(t)}$ is the percentage of germinated spores at time t, P_{max} (%) is the asymptotic P value at $t \rightarrow +\infty$, μ_G (%.h^{-1}) is the slope term of the tangent line through the inflection point and λ_G (h) is the geometrical latency (t-axis intercept of the tangent through the inflection point)	[ABE 99, JUD 08, GOU 12]
	Logistic function $$P_{(t)} = \frac{P_{max}}{1+exp[k(\tau-t)]}$$ [3.4] where t (h), $P_{(t)}$ and P_{max} have the same definitions as above, τ (h) is the inflection point where P=$Pmax/2$ (with τ equal to the germination time t_G), and k (%.h−1) is related to the slope of the tangent line through the inflection point. By deriving, the latency prior to the germination λ_G (h) is extrapolated	[DAN 05b, SAM 07a, JUD 08]
	Asymmetric model $$P_{(t)} = P_{max}\left[1 - \frac{1}{1+\left(\frac{t}{\tau}\right)^d}\right]$$ [3.5] In addition to the parameters previously defined, d is a shape parameter. By deriving, the latency prior to the germination λ_G (h) is extrapolated	[DAN 11]
Mycelium growth	Baranyi model (See equation in Chapter 1)	[MEM 00, MEM 01, VAL 01, SAM 05, GAR 10, SAM 10, GAR 11, AST 12]
	Two-phase linear $$D_{(t)} = \begin{cases} D_0, & t \leq \lambda \\ D_0 + \mu.(t - \lambda), & t > \lambda \end{cases}$$ [3.6] where t is the time (day), $D_{(t)}$ (mm) is the colony diameter at time t, D_0 (mm) is the initial colony diameter, λ is the "latency" prior to the increase in the colony diameter (h) and μ (mm.h^{-1}) is the mycelium radial growth rate	[GOU 10, GOU 11, MAU 11, HUC 13]

Table 3.3. *Main primary models applied to describe germination or mycelium proliferation. Adapted from [DAG 13]*

Concerning the secondary models, there is a long list of models for mould spoilage applications (see the review made by Dagnas [DAG 13]). Among them, the cardinal value models, plugged into a Gamma structure [3.2], have gained interest. In Table 3.4, the main factors studied through this type of model are presented:

$$\mu = \mu opt \times \gamma(a_w) \times \gamma(T) \times \gamma(pH) \times \gamma(\ldots) \qquad [3.2]$$

In [3.2], each gamma term is parameterized such as its value varies between 0 and 1. At 0, the inhibition due to this factor is maximal, at 1, the inhibition is null. In the absence of inhibition, the mycelium growth rate of mould is at its maximal value, µopt.

Mold growth	Parameter	Factor	References
Germination	t_G[a]	Ethanol	[DAN 05b]
	λ_G[b], μ_G[c]	T	[GOU 12]
Mycelium growth	μ[d]	Ph	[MEM 01]
	M	a_w	[ROS 01, SAU 01]
	M	T	[PAR 04]
	M	Ethanol	[DAN 05a]
	M	a_w, T	[AST 12, GAR 11, JUD 10, PAN 10, TAS 07, HUC 13]
	M	a_w, T, pH	[MAU 11, NEV 09, PAN 03]
	λ[e], μ	a_w	[MAR 09]
	λ, μ	a_w, T, pH	[DAG 14]
	λ, μ	T	[GOU 11, GOU 10]
	Λ	Acetic and propionic acid	[DAG 15]

[a] t_G, germination time: time to get 50% viable spores germinated (h)
[b] λ_G, latency prior to the germ tube formation during the germination (h)
[c] μ_G, germination rate ($\%.h^{-1}$)
[d] μ, mycelium growth rate ($mm.day^{-1}$)
[e] λ, "latency" prior to the mycelium growth (h)

Table 3.4. *List of factors studied with secondary models based on cardinal values, either in germination or mycelium proliferation*

The reason of an increasing interest for cardinal value models plugged into a Gamma structure is relatively simple and could be summarized in three words: flexibility, interpretation and parsimony:

– flexibility: the multiplication form of the structure enables us to add later on the effect of an extra factor without modifying the existing model;

– interpretation: each Gamma term is described with a cardinal value model (containing cardinal parameters such as Tmin for the temperature factor, pHmin for the pH factor, minimum inhibitory concentration (MIC) for acids, etc.) which could be easily interpreted by food microbiologists; that could also be a valuable help in defining prior distribution when a Bayesian approach is chosen to estimate the parameters;

– parsimony: cardinal value models describe the effect of the factor with a limited number of parameters (e.g. 3 for the temperature and 2 for the acids).

3.3.4. *Use of models to derive management options which enable us to prevent/control food spoilage*

Once primary and secondary models have been developed, they have to be incorporated into a more comprehensive framework in order to be used for recommendations of management options to prevent/control food spoilage. At this stage, it is important to determine, for each given food product application, the formulation and process window (i.e. limit of acceptable changes). This determination has to take into consideration the criteria related to the technological feasibility, the legislation, the local or regional requirement associated with a specific food, the consumer's preference, etc. With bakery products, a_w is the key factor influencing mould growth [DAG 14, GUY 05, HUC 13, ROS 01]. However, a_w cannot be changed too much without altering the product organoleptic properties. The formulation window is then rather narrow. Another option is to set the BBD in a way that there is no development of visible mycelium before consumption. For most bakery products, this is an interesting option to consider, knowing that these products have relatively long shelf-life (several weeks).

Below, two frameworks aiming at illustrating how predictive models could be used in a_w/BBD determination for bakery-products are detailed. But before moving onto the details of the study, let us introduce the variable time before consumption (TBC). TBC is a time, i.e. a duration, it corresponds to the actual shelf-life of the product. It is then different from the BBD, which is a date. BBD is a fixed value while TBC is a variable value as not all consumers are going to eat the product at its BBD.

Next, let us define the time to visible growth (tv): it corresponds to the sum of lag time plus the visible limit divided by the growth rate [3.7]:

$$t_v = \lambda + D_v / \mu \quad [3.7]$$

In [3.7], λ is the lag time (days), μ is the radial growth rate (mm.day-1) and Dv is the visible diameter (mm). Dv is generally fixed in a range from 1 [HUC 13] to 3 mm [GOU 10]. λ and μ are directly derived from secondary models (Table 3.4).

A first framework, followed, for instance, by [HUC 13], consists of considering that TBC matches BBD and then, that TBC equals tv. In other words, the BBD is set such as the growth is visible at the BBD, for 100% products. Huchet *et al.* [HUC 13] have predicted BBD for various a_w and temperature values. For instance, at a temperature of 20°C, with an a_w of 0.83, they estimated a BBD equaled to 25 days (rigorously, we should say "BBD set to have a period of 25 days between product factory release and BBD").

This approach is deterministic: the uncertainty due to model error is not taken into account, the temperature is a fixed value and more importantly, it is assumed that all consumers are going to eat the product at its BBD. Thus, this approach, although very straightforward, cannot be recommended for BBD determination as it includes a conservative assumption. Nevertheless, it could be used to predict the relative influence of various management options.

Building on this, and considering the advantage of having a probabilistic approach (see Chapter 2), a second framework could be considered:

– the storage temperature is incorporated in the model as a range of values, for instance to consider that consumer home temperature may vary between consumers, and also, with the season;

– the model error is included in the model, this has to be done following the model fitting hypothesis. For instance, in the case of a model developed on the square root of μ, $\sqrt{\mu}$ has to be estimated using a normal distribution: $\sqrt{\mu} \sim N(\widehat{\sqrt{\mu}}, sd^2)$, in which the model error is incorporated through the variance, sd^2 (sd = root mean square error);

– the TBC is chosen to reflect the variability in consumers' preference and habits, and no longer set at a single value.

In Table 3.5, an example of probabilistic variables describing storage temperature, model error and TBC is provided. This example is given only for illustration purposes, even if it is based upon realistic information, with an application on a Brioche-type product.

Factors	Assumptions	Mathematical description
Storage temperature	Split the temperature into two periods over the year: (1) a "cold" period representing autumn and winter where indoor temperatures are assumed to be around 19°C, with a minimum and a maximum of 16 and 22°C, respectively; and (2) a "warm" period representing spring and summer where indoor temperatures are assumed to vary between 18 and 30°C with a likely value of 21°C	Season=Bernoulli(0.5) T_{winter}=Pert(16,19,22) T_{summer}=Pert(18,21,30) Temperature=Season x T_{winter} + (1-Season) x T_{summer}
Growth rate and lag	With a model developed on the square root, the response is fitted by a normal distribution using $\sqrt{\mu}$ and $\sqrt{1/lag}$ as responses	$\sqrt{\mu} \sim N(\widehat{\sqrt{\mu}}, sd_{\mu}^2)$ $\widehat{\sqrt{\mu}}$: expected value, output of predictive model sd_{μ}: root mean square error $\mu = \sqrt{\mu}^2$ $\sqrt{1/lag} \sim N(\widehat{\sqrt{1/lag}}, sd_{lag}^2)$ $\widehat{\sqrt{1/lag}}$: expected value, output of predictive model sd_{lag}: root mean square error $Lag = 1/\sqrt{1/lag}^2$
Time before consumption (TBC)	Consumer likely eats the product following its purchase. For example, 35% product eaten in the first quarter of BBD, 35% in the second quarter, 20% in the third and fourth quarter, and 10% over the BBD	Time=(Uniform(1,BBD/4), Uniform(BBD/4, BBD/2), Uniform(BBD/2,BBD), Uniform(BBD, BBDx1.2)) Ratio=(0.35;0.35;0.2;0.1) TBC=Discrete(Ratio;Time)

Table 3.5. *Description of probabilistic inputs included in a model aiming at estimating the mould spoilage rate of bakery product*

Using these probabilistic variables on one hand and predictive models (Tables 3.3 and 3.4) on other hand, it is possible to calculate the time for visible growth, tv [3.7] and then to calculate a spoilage rate, defined as the probability that the product is consumed after the time of visible mould growth [3.8]:

$$SR = Pr(t_v \leq TBC) \quad [3.8]$$

To go a step further in the illustration, let us take the models developed by Dagnas *et al.* [DAG 15, DAG 14] on the effect of temperature, a_w, acetic and propionic acids to calculate the spoilage rate at the time of consumption. The authors have worked with *Eurotiumrepens*, *Penicilliumcorylophilum* and *Aspergillus niger*, three mould species frequently encountered in bakery products. If we consider that the spoilage is due to one of the three species (but never two or three at the same time), assuming a scarce in-factory contamination, the overall spoilage rate is calculated as follows:

$$S_R = \frac{[Pr(t_{v\,Aniger} \leq TBC) + Pr(t_{v\,Erepens} \leq TBC) + Pr(t_{v\,Pcorylophilum} \leq TBC)]}{3} \quad [3.9]$$

In [3.9], each term $Pr(t_{v\ \text{Mould species}} \leq TBC)$ is estimated using predictive models built on μ and λ (including error terms), fixed inputs (a_w, pH, acetic and propionic acid concentration, BBD and Dv) and probabilistic variables (defined as indicated in Table 3.5). Note that TBC is a probabilistic variable deduced from BBD, and then not strictly speaking an input (TBC is a latent variable).

In this illustration, a brioche (sourdough-based product) is taken as an example. Considering that the product is slightly acidic (pH 5.2) and contains 5 $mmol.kg^{-1}$ of propionic acid and 17 $mmol.kg^{-1}$ of acetic acid [ZHA 10], it is estimated that after 20 days, the spoilage rate of a brioche with an a_w of 0.86 is equaled to $5.\ 10^{-2}$ (one contaminated product in 20 spoiled). This is a high, but realistic value considering that in this illustration, the product is initially contaminated, i.e. contaminated in the factory, before packaging. The model framework also enables us to identify combinations of a_w and BBD which lead to equivalent spoilage rate. For instance, a BBD of 25 days combined with an a_w of 0.852 is equivalent in terms of spoilage rate to a BBD of 35 days combined with an a_w of 0.837. More iso-probability combinations are depicted in Figure 3.3.

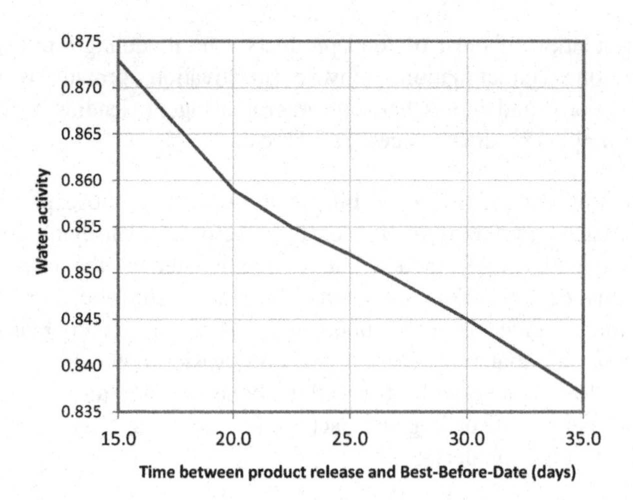

Figure 3.3. *Combinations of water activity and best-before-date leading to an equaled spoilage rate. Example of a sourdough-based bakery product, such as a brioche (pH=5.2)*

3.4. Conclusion

In this chapter, after a short introduction on microbial food spoilage, two applications of food spoilage modeling are presented. The first application concerns heat-treated food products spoiled by spore-forming bacteria, the second application concerns intermediate a_w products spoiled by mould. In the first case, a relatively complex probabilistic model aiming at estimating the commercial sterility failure rate due to *G. stearothermophilus* has been presented. The model was built with a separation of uncertainty and variability and its interpretation through scenario analysis. Concerning mould spoilage, an example of a probabilistic model developed to help in formulation design and BBD determination of bakery products has been introduced.

These two examples illustrate the diversity of applications. They emphasize as well that the mathematical structure of the predictive models used to quantify microbial food spoilage is relatively similar to what has been developed in the microbial modeling area for more than 30 years (Chapter 1).

However, the rationale behind modeling for food spoilage or food safety application is relatively different. Indeed, it is difficult, not to say impossible, to eradicate spoiling microorganisms since they are well adapted to process and/or formulation of most of food products. The modeling work involves more than one contamination pathway (inactivation, growth as well as recontamination), and more than one microbial agent, leading to building relatively complex modular process risk models.

Due to this complexity, probabilistic tools such as those presented in Chapter 2 are particularly valuable; they help in identifying and then evaluating quantitatively (in absolute or relative terms) the efficiency of various management options to control the risk of spoilage. Note that, as spoiling microorganisms are abundantly spread in food manufacture (environment, raw materials, etc.), it is slightly easier to benchmark a model output built for spoilage with observations done in a day-to-day production run than in the case of pathogenic bacteria (where the spoilage rate and the acceptable limit are smaller).

3.5. Bibliography

[ABE 99] ABELLANA M., BENEDI J., SANCHIS V. *et al.*, "Water activity and temperature effects on germination and growth of *Eurotium amstelodami*, *E. chevalieri* and *E. herbariorum* isolates from bakery products", *Journal of Applied Microbiology,* vol. 3, pp. 371–380, 1999.

[AND 11] ANDERSON N.M., LARKIN J.W., COLE M.B. *et al.*, "Food safety objective approach for controlling *Clostridium botulinum* growth and toxin production in commercially sterile foods", *Journal of Food Protection,* vol. 11, pp. 1956–1989, 2011.

[AND 13] ANDRE S., ZUBER F., REMIZE F., "Thermophilic spore-forming bacteria isolated from spoiled canned food and their heat resistance. Results of a French ten-year survey", *International Journal of Food Microbiology,* vol. 2, pp. 134–143, 2013.

[AST 12] ASTORECA A., VAAMONDE G., DALCERO A. *et al.*, "Modelling the effect of temperature and water activity of *Aspergillus flavus* isolates from corn", *International Journal of Food Microbiology*, 2012.

[BAT 02] BATTEY A.S., DUFFY S., SCHAFFNER D.W., "Modeling yeast spoilage in cold-filled ready-to-drink beverages with *Saccharomyces cerevisiae*, *Zygosaccharomyces bailii*, and *Candida lipolytica*", *Applied and Environmental Microbiology,* vol. 4, pp. 1901–1906, 2002.

[BIG 21] BIGELOW W.D., "The logarithmic nature of thermal death time curves", *Journal of Infectious Diseases,* vol. 5, pp. 528–236, 1921.

[BLA 06] BLACKBURN C.D.W., *Food Spoilage Microorganisms*, Elsevier, London, UK, 2006.

[BRU 07] BRUL S., VAN GERWEN S., ZWIETERING M., *Modelling Microorganisms in Food*, CRC Press, New York, 2007.

[CER 01] CERF O., DAVEY K.R., "An explanation of non-sterile (leaky) milk packs in well-operated UHT plant", *Food and Bioproducts Processing*, vol. 4, pp. 219–222, 2001.

[COD 93] CODEX ALIMENTARIUS COMMISSION, Code of hygienic practice for low-acid and acidified low-acid canned foods, CAC/RCP 23-1979, 1993.

[DAG 13] DAGNAS S., MEMBRÉ J.M., "Predicting and preventing mold spoilage of food products", *Journal of Food Protection*, vol. 3, pp. 538–551, 2013.

[DAG 14] DAGNAS S., ONNO B., MEMBRÉ J.-M., "Modeling growth of three bakery product spoilage molds as a function of water activity, temperature and pH", *International Journal of Food Microbiology*, vol. 0, pp. 95–104, 2014.

[DAG 15] DAGNAS S., GAUVRY E., ONNO B. *et al.*, "Quantifying effect of lactic, acetic and propionic acids on growth of molds isolated from spoiled bakery products", *Journal of Food Protection*, 2015.

[DAN 05a] DANTIGNY P., GUILMART A., RADOI A. *et al.*, "Modelling the effect of ethanol on growth rate of food spoilage moulds", *International Journal of Food Microbiology*, vol. 3, pp. 261–269, 2005.

[DAN 05b] DANTIGNY P., TCHOBANOV I., BENSOUSSAN M. *et al.*, "Modeling the effect of ethanol vapor on the germination time of *Penicillium chrysogenum*", *Journal of Food Protection*, vol. 6, pp. 1203–1207, 2005.

[DAN 11] DANTIGNY P., NANGUY S.P.M., JUDET-CORREIA D. *et al.*, "A new model for germination of fungi", *International Journal of Food Microbiology*, vol. 2, pp. 176–181, 2011.

[DAN 13] DANTIGNY P., PANAGOU E.Z., *Predictive Mycology*, Nova Science Publishers, Inc., New York, 2013.

[DAO 10] DAO T., DEJARDIN J., BENSOUSSAN M. *et al.*, "Use of the Weibull model to describe inactivation of dry-harvested conidia of different *Penicillium* species by ethanol vapours", *Journal of Applied Microbiology*, vol. 2, pp. 408–414, 2010.

[DAO 11] DAO T., DANTIGNY P., "Control of food spoilage fungi by ethanol", *Food Control*, vol. 3–4, pp. 360–368, 2011.

[DEN 81] DENNY C.B., "Thermophilic organisms involved in food spoilage: introduction", *Journal of Food Protection,* vol. 2, pp. 144–145, 1981.

[DEN 03] DEN AANTREKKER E.D., BOOM R.M., ZWIETERING M.H. *et al.*, "Quantifying recontamination through factory environments – a review", *International Journal of Food Microbiology,* vol. 2, pp. 117–130, 2003.

[DES 15] DESCHUYFFELEER N., VERMEULEN A., DAELMAN J. *et al.*, "Modelling of the growth/no growth interface of *Wallemia sebi* and *Eurotium herbariorum* as a function of pH, aw and ethanol concentration", *International Journal of Food Microbiology*, pp. 77–85, 2015.

[DOM 07] DOMINGUEZ S.A., SCHAFFNER D.W., "Development and validation of a mathematical model to describe the growth of *Pseudomonas* spp. in raw poultry stored under aerobic conditions", *International Journal of Food Microbiology,* vol. 3, pp. 287–295, 2007.

[EUR 02] EUROPEAN COMMISSION (EC), "REGULATION (EC) No 178/2002 of the European parliament and of the Council of 28 January 2002 laying down the general principles and requirements of food law, establishing the European Food Safety Authority and laying down procedures in matters of food safety", *Official Journal of the European Union*, pp. 1–24, 2002.

[FAN 13] FANG T., LIU Y., HUANG L., "Growth kinetics of *Listeria monocytogenes* and spoilage microorganisms in fresh-cut cantaloupe", *Food Microbiology,* vol. 1, pp. 174–181, 2013.

[FAO 04] FAO/WHO, "Risk assessment of *Listeria monocytogenes* in ready-to-eat foods: interpretative summary", in FAO (ed.), *Microbiological Risk Assessment Series,* Rome, no. 4, p. 78, 2004.

[GAR 10] GARCIA D., RAMOS A.J., SANCHIS V. *et al.*, "Modelling mould growth under suboptimal environmental conditions and inoculum size", *Food Microbiology,* vol. 7, pp. 909–917, 2010.

[GAR 11] GARCIA D., RAMOS A.J., SANCHIS V. *et al.*, "Modelling the effect of temperature and water activity in the growth boundaries of *Aspergillus ochraceus* and *Aspergillus parasiticus*", *Food Microbiology,* vol. 3, pp. 406–417, 2011.

[GHA 02] GHANI A.G.A., FARID M.M., CHEN X.D., "Theoretical and experimental investigation of the thermal inactivation of *Bacillus stearothermophilus* in food pouches", *Journal of Food Engineering,* vol. 3, pp. 221–228, 2002.

[GIM 04] GIMENEZ B., DALGAARD P., "Modelling and predicting the simultaneous growth of *Listeria monocytogenes* and spoilage micro-organisms in cold-smoked salmon", *Journal of Applied Microbiology,* vol. 1, pp. 96–109, 2004.

[GOU 10] GOUGOULI M., KOUTSOUMANIS K.P., "Modelling growth of Penicillium expansum and Aspergillus niger at constant and fluctuating temperature conditions", *International Journal of Food Microbiology*, vol. 2–3, pp. 254–262, 2010.

[GOU 11] GOUGOULI M., KALANTZI K., BELETSIOTIS E. *et al.*, "Development and application of predictive models for fungal growth as tools to improve quality control in yogurt production", *Food Microbiology*, vol. 8, pp. 1453–1462, 2011.

[GOU 12] GOUGOULI M., KOUTSOUMANIS K.P., "Modeling germination of fungal spores at constant and fluctuating temperature conditions", *International Journal of Food Microbiology*, vol. 3, pp. 153–161, 2012.

[GUS 11] GUSTAVSSON J., CEDERBERG C., SONESSON U. *et al.*, Rome, Italy, FAO, 2011.

[GUY 02] GUYNOT M.E., RAMOS A.J., SALA D. *et al.*, "Combined effects of weak acid preservatives, pH and water activity on growth of *Eurotium* species on a sponge cake", *International Journal of Food Microbiology*, vol. 1–2, pp. 39–46, 2002.

[GUY 05] GUYNOT M.E., MARIN S., SANCHIS V. *et al.*, "An attempt to optimize potassium sorbate use to preserve low pH (4.5–5.5) intermediate moisture bakery products by modelling *Eurotium* spp., *Aspergillus* spp. and *Penicillium corylophilum* growth", *International Journal of Food Microbiology*, pp. 169–177, 2005.

[HAA 14] HAAS C.N., ROSE J.B., GERBA C.P., *Quantitative Microbial Risk Assessment*, John Wiley & Sons, Inc, Hoboken, New Jersey, 2014.

[HOR 73] HORNER K.J., ANAGNOSTOPOULOS G.D., "Combined effects of water activity, pH and temperature on the growth and spoilage potential of fungi", *Journal of Applied Microbiology*, vol. 3, pp. 427–436, 1973.

[HUC 13] HUCHET V., PAVAN S., LOCHARDET A. *et al.*, "Development and application of a predictive model of *Aspergillus candidus* growth as a tool to improve shelf life of bakery products", *Food Microbiology*, vol. 2, pp. 254–259, 2013.

[HUI 96] HUIS IN'T VELD J.H.J., "Microbial and biochemical spoilage of foods: an overview", *International Journal of Food Microbiology*, vol. 1, pp. 1–18, 1996.

[ITO 81] ITO K.A., "Thermophilic organisms in food spoilage: flat-sour aerobes", *Journal of Food Protection*, vol. 2, pp. 157–163, 1981.

[JAG 03] JAGANNATH A., TSUCHIDO T., "Validation of a polynomial regression model: the thermal inactivation of Bacillus subtilis spores in milk", *Letters in Applied Microbiology*, vol. 5, pp. 399–404, 2003.

[JUD 08] JUDET D., BENSOUSSAN M., PERRIER-CORNET J.-M. *et al.*, "Distributions of the growth rate of the germ tubes and germination time of *Penicillium chrysogenum* conidia depend on water activity", *Food Microbiology*, vol. 7, pp. 902–907, 2008.

[JUD 10] JUDET-CORREIA D., BOLLAERT S., DUQUENNE A. *et al.*, "Validation of a predictive model for the growth of *Botrytis cinerea* and *Penicillium expansum* on grape berries", *International Journal of Food Microbiology,* vol. 1–2, pp. 106–113, 2010.

[KOT 97] KOTZEKIDOU P., "Heat resistance of *Byssochlamys nivea, Byssochlamys fulva* and *Neosartorya fischeri* isolated from canned tomato paste", *Journal of Food Science,* vol. 2, pp. 410–412, 1997.

[KOU 06] KOUTSOUMANIS K., STAMATIOU A., SKANDAMIS P. *et al.*, "Development of a microbial model for the combined effect of temperature and ph on spoilage of ground meat, and validation of the model under dynamic temperature conditions", *Applied and Environmental Microbiology*, pp. 124–134, 2006.

[KOU 09] KOUTSOUMANIS K., "Modeling food spoilage in microbial risk assessment", *Journal of Food Protection,* vol. 2, pp. 425–427, 2009.

[LAH 05] LAHLALI R., SERRHINI M.N., JIJAKLI M.H., "Studying and modelling the combined effect of temperature and water activity on the growth rate of *P. expansum*", *International Journal of Food Microbiology,* vol. 3, pp. 315–322, 2005.

[LER 12] LEROI F., FALL P.A., PILET M.F. *et al.*, "Influence of temperature, pH and NaCl concentration on the maximal growth rate of *Brochothrix thermosphacta* and a bioprotective bacteria Lactococcus piscium CNCM I-4031", *Food Microbiology,* vol. 2, pp. 222–228, 2012.

[MAN 11] MANTOAN D., SPILIMBERGO S., "Mathematical modeling of yeast inactivation of freshly squeezed apple juice under high-pressure carbon dioxide", *Critical Reviews in Food Science and Nutrition,* vol. 1, pp. 91–97, 2011.

[MAR 09] MARIN S., COLOM C., SANCHIS V. *et al.*, "Modelling of growth of aflatoxigenic *A. flavus* isolates from red chilli powder as a function of water availability", *International Journal of Food Microbiology,* vol. 3, pp. 491–496, 2009.

[MAR 15] MARVIG C.L., KRISTIANSEN R.M., NIELSEN D.S., "Growth/no growth models for *Zygosaccharomyces rouxii* associated with acidic, sweet intermediate moisture food products", *International Journal of Food Microbiology*, pp. 51–57, 2015.

[MAU 11] MAURICE S., COROLLER L., DEBAETS S. *et al.*, "Modelling the effect of temperature, water activity and pH on the growth of *Serpula lacrymans*", *Journal of Applied Microbiology,* vol. 6, pp. 1436–1446, 2011.

[MCK 04] MCKELLAR R.C., LU X., *Modeling Microbial Responses in Food,* CRC Press, London, 2004.

[MEJ 13] MEJLHOLM O., DALGAARD P., "Development and validation of an extensive growth and growth boundary model for psychrotolerant *Lactobacillus* spp. in seafood and meat products", *International Journal of Food Microbiology*, vol. 2, pp. 244–260, 2013.

[MEJ 15] MEJLHOLM O., DALGAARD P., "Modelling and predicting the simultaneous growth of *Listeria monocytogenes* and psychrotolerant lactic acid bacteria in processed seafood and mayonnaise-based seafood salads", *Food Microbiology*, pp. 1–14, 2015.

[MEM 00] MEMBRÉ J.M., KUBACZKA M., "Predictive modelling approach applied to spoilage fungi: growth of *Penicillium brevicompactum* on solid media", *Letters in Applied Microbiology*, vol. 3, pp. 247–250, 2000.

[MEM 01] MEMBRÉ J.-M., KUBACZKA M., CHENE C., "Growth rate and growth no-growth interface of *Penicillium brevicompactum* as functions of pH and preservative acids", *Food Microbiology*, vol. 5, pp. 531–538, 2001.

[MEM 07] MEMBRÉ J.M., BASSETT J., GORRIS L.G.M., "Applying the food safety objective and related standards to thermal inactivation of Salmonella in poultry meat", *Journal of Food Protection*, vol. 9, pp. 2036–2044, 2007.

[MEM 11] MEMBRÉ J.-M., VAN ZUIJLEN A., "A probabilistic approach to determine thermal process setting parameters: application for commercial sterility of products", *International Journal of Food Microbiology*, pp. 413–420, 2011.

[MEM 12] MEMBRÉ J.M., "Setting of thermal processes in a context of food safety objectives (FSOs) and related concepts", in VALDRAMIDIS V.J.F.M., VAN I. (eds), *Progress on Quantitative Approaches of Thermal Food Processing*, Nova Science Publishers, New York, 2012.

[MOK 05] MOKHTARI A., FREY H.C., "Sensitivity analysis of a two-dimensional probabilistic risk assessment model using analysis of variance", *Risk Analysis*, vol. 6, pp. 1511–1529, 2005.

[NAU 00] NAUTA M.J. "Separation of uncertainty and variability in quantitative microbial risk assessment models", *International Journal of Food Microbiology*, vol. 57, nos. 1–2, pp. 9–18, 2000.

[NAU 01] NAUTA M.J., "A modular process risk model structure for quantitative microbiological risk assessment and its application in an exposure assessment of *Bacillus cereus* in a REPFED", *Rivm*, p. 100, 2001.

[NEV 09] NEVAREZ L., VASSEUR V., LE MADEC A. *et al.*, "Physiological traits of *Penicillium glabrum* strain LCP 08.5568, a filamentous fungus isolated from bottled aromatised mineral water", *International Journal of Food Microbiology*, vol. 3, pp. 166–171, 2009.

[PAN 03] PANAGOU E.Z., SKANDAMIS P.N., NYCHAS G.J.E., "Modelling the combined effect of temperature, pH and aw on the growth rate of *Monascus ruber*, a heat-resistant fungus isolated from green table olives", *Journal of Applied Microbiology,* vol. 1, pp. 146–156, 2003.

[PAN 10] PANAGOU E.Z., CHELONAS S., CHATZIPAVLIDIS I. *et al.*, "Modelling the effect of temperature and water activity on the growth rate and growth/no growth interface of *Byssochlamys fulva* and *Byssochlamys nivea*", *Food Microbiology,* vol. 5, pp. 618–627, 2010.

[PAR 04] PARRA R., MAGAN N., "Modelling the effect of temperature and water activity on growth of *Aspergillus niger* strains and applications for food spoilage moulds", *Journal of Applied Microbiology,* vol. 2, pp. 429–438, 2004.

[PIT 75] PITT J.I., "Xerophilic fungi and the spoilage of foods of plant origin", in DUCKWORTH R.B. (ed.), *Water Relations of Food,* Academic Press, London, 1975.

[PIT 77] PITT J.I., HOCKING A.D., "Influence of solute and hydrogen ion concentration on the water relations of some xerophilic fungi", *Journal of General Microbiology,* vol. 1, pp. 35–40, 1977.

[POU 10] POUILLOT R., DELIGNETTE-MULLER M.L., "Evaluating variability and uncertainty separately in microbial quantitative risk assessment using two R packages", *International Journal of Food Microbiology,* vol. 3, pp. 330–340, 2010.

[POU 15] POUILLOT R., CHEN Y., HOELZER K., "Modeling number of bacteria per food unit in comparison to bacterial concentration in quantitative risk assessment: impact on risk estimates", *Food Microbiology,* PB, pp. 245–253, 2015.

[PRE 10] PREVOST S., ANDRE S., REMIZE F., "PCR detection of thermophilic spore-forming bacteria involved in canned food spoilage", *Current Microbiology,* vol. 6, pp. 525–533, 2010.

[PUJ 15a] PUJOL L., ALBERT I., MAGRAS C. *et al.*, "Estimation and evaluation of management options to control and/or reduce the risk of not complying with commercial sterility", *International Journal of Food Microbiology*, 2015.

[PUJ 15b] PUJOL L., ALBERT I., MAGRAS C. *et al.*, "Probabilistic exposure assessment model to estimate aseptic UHT product failure rate", *International Journal of Food Microbiology*, pp. 124–141, 2015.

[PUJ 15c] PUJOL L., JOHNSON N.B., MAGRAS C. *et al.*, "Added value of experts' knowledge to improve a quantitative microbial exposure assessment model – application to aseptic-UHT food products", *International Journal of Food Microbiology,* 2015.

[RIG 13] RIGAUX C., DENIS J.-B., ALBERT I. *et al.*, "A meta-analysis accounting for sources of variability to estimate heat resistance reference parameters of bacteria using hierarchical Bayesian modeling: estimation of D at 121.1°C and pH 7, zT and zpH of *Geobacillus stearothermophilus*", *International Journal of Food Microbiology,* vol. 2, pp. 112–120, 2013.

[RIG 14] RIGAUX C., ANDRE S., ALBERT I. *et al.*, "Quantitative assessment of the risk of microbial spoilage in foods. Prediction of non-stability at 55°C caused by *Geobacillus stearothermophilus* in canned green beans", *International Journal of Food Microbiology*, pp. 119–128, 2014.

[ROS 01] ROSSO L., ROBINSON T.P., "A cardinal model to describe the effect of water activity on the growth of moulds", *International Journal of Food Microbiology,* vol. 3, pp. 265–273, 2001.

[SAL 02] SALTELLI A., "Sensitivity analysis for importance assessment", *Risk Analysis,* vol. 3, pp. 579–590, 2002.

[SAM 05] SAMAPUNDO S., DEVLIEGHERE F., DE MEULENAER B. *et al.*, "Predictive modelling of the individual and combined effect of water activity and temperature on the radial growth of *Fusarium verticilliodes* and *F. proliferatum* on corn", *International Journal of Food Microbiology,* vol. 1, pp. 35–52, 2005.

[SAM 07a] SAMAPUNDO S., DEVLIEGHERE F., DE MEULENAER B. *et al.*, "Growth kinetics of cultures from single spores of *Aspergillus flavus* and *Fusarium verticillioides* on yellow dent corn meal", *Food Microbiology,* vol. 4, pp. 336–345, 2007.

[SAM 07b] SAMAPUNDO S., DEVLIEGHERE F., GEERAERD A.H. *et al.*, "Modelling of the individual and combined effects of water activity and temperature on the radial growth of *Aspergillus flavus* and *A. parasiticus* on corn", *Food Microbiology,* vol. 5, pp. 517–529, 2007.

[SAM 10] SAMAPUNDO S., DESCHUYFFELEER N., VAN LAERE D. *et al.*, "Effect of NaCl reduction and replacement on the growth of fungi important to the spoilage of bread", *Food Microbiology,* vol. 6, pp. 749–756, 2010.

[SAN 09] SANT'ANA A.S., ROSENTHAL A., MASSAGUER P.R., "Heat resistance and the effects of continuous pasteurization on the inactivation of *Byssochlamys fulva* ascospores in clarified apple juice", *Journal of Applied Microbiology,* vol. 1, pp. 197–209, 2009.

[SAU 01] SAUTOUR M., DANTIGNY P., DIVIES C. *et al.*, "A temperature-type model for describing the relationship between fungal growth and water activity", *International Journal of Food Microbiology,* vol. 1–2, pp. 63–69, 2001.

[SME 14] SMELT J.P.P.M., BRUL S., "Thermal inactivation of microorganisms", *Critical Reviews in Food Science and Nutrition,* vol. 10, pp. 1371–1385, 2014.

[STA 15] STAHL V., NDOYE F.T., EL JABRI M. *et al.*, "Safety and quality assessment of ready-to-eat pork products in the cold chain", *Journal of Food Engineering*, pp. 43–52, 2015.

[STR 09] STRATFORD M., PLUMRIDGE A., NEBE-VON-CARON G. *et al.*, "Inhibition of spoilage mould conidia by acetic acid and sorbic acid involves different modes of action, requiring modification of the classical weak-acid theory", *International Journal of Food Microbiology,* vol. 1, pp. 37–43, 2009.

[STU 73] STUMBO C.R., *Thermobacteriology in Food Processing,* Academic Press, Inc., New York, 1973.

[TAS 07] TASSOU C.C., PANAGOU E.Z., NATSKOULIS P. *et al.*, "Modelling the effect of temperature and water activity on the growth of two ochratoxigenic strains of *Aspergillus carbonarius* from Greek wine grapes", *Journal of Applied Microbiology,* vol. 6, pp. 2267–2276, 2007.

[THO 02] THOMPSON K.M., "Variability and uncertainty meet risk management and risk communication", *Risk Analysis*, vol. 22, pp. 647–654, 2002.

[UYT 09] UYTTENDAELE M., BAERT K., GRIJSPEERDT K. *et al.*, "Comparing the effect of various contamination levels for Salmonella in chicken meat preparations on the probability of illness in Belgium", *Journal of Food Protection,* vol. 10, pp. 2093–2105, 2009.

[VAL 01] VALIK L., PIECKOVA E., "Growth modelling of heat-resistant fungi: the effect of water activity", *International Journal of Food Microbiology,* vol. 1–2, pp. 11–17, 2001.

[VAN 98] VAN GERWEN S.J.C., ZWIETERING M.H., "Growth and inactivation models to be used in quantitative risk assessments", *Journal of Food Protection,* vol. 11, pp. 1541–1549, 1998.

[ZHA 10] ZHANG C., BRANDT M.J., SCHWAB C. *et al.*, "Propionic acid production by cofermentation of *Lactobacillus buchneri* and *Lactobacillus diolivorans* in sourdough", *Food Microbiology,* vol. 3, pp. 390–395, 2010.

[ZIM 13] ZIMMERMANN M., SCHAFFNER D.W., ARAGÃO G.M.F., "Modeling the inactivation kinetics of *Bacillus coagulans* spores in tomato pulp from the combined effect of high pressure and moderate temperature", *LWT – Food Science and Technology,* vol. 1, pp. 107–112, 2013.

[ZWI 02] ZWIETERING M.H., "Quantification of microbial quality and safety in minimally processed foods", *International Dairy Journal,* vol. 2–3, pp. 263–271, 2002.

4

Modeling Microbial Responses: Application to Food Safety

4.1. Introduction

In the last few decades, the incidence of foodborne diseases has increased despite the introduction of HACCP and the proliferation of food safety regulations. The increased incidence of foodborne diseases, caused by changes in agricultural and food processing practices, increasing international trade and social changes, stresses the need for more effective food quality and safety assurance systems. Current approaches to food safety that rely heavily on regulatory inspection and sampling regimes cannot sufficiently guarantee consumer protection since they are time-consuming, the 100% inspection and sampling is financially and logistically impossible [ARM 97], and temperature abuses, a major cause of safety problems especially for products that are stored at chill conditions, cannot be controlled and often deviate from specifications [KOU 15, KOU 10].

An alternative approach to traditional methods of safety assurance is to use quantitative microbiological tools [MCM 97]. Quantitative microbiology seems as an attractive and pertinent tool for food safety management. Advancement of quantitative microbiology has allowed the significant progress toward effective, validated modeling of food safety in the last years, while significant effort for developing new predictive microbiology tools is still in progress. Deterministic models, models that provide point estimates of microbial populations, have been recognized as being incompetent to take into account the important sources of variability, and as such, they have been questioned with regard to

---

Chapter written by Maria GOUGOULI and Konstantinos KOUTSOUMANIS.

their value in microbial risk assessment (MRA) and food safety management [JUN 03, KOS 11, NIC 96, POS 03]. Such a deficiency stressed the need for the development of models capable of incorporating the variation of model parameters, and motivated the commencement of more sophisticated modeling approach called probabilistic or stochastic modeling. Probabilistic modeling is being used with an increasing frequency in the area of food safety and it has been extensively applied in quantitative MRAs [CAS 98, FAO 04, FDA 03], in quality and safety management systems [GIA 01, KOU 02, KOU 05] and recently for more specific topics, such as the evaluation of the effects of food processing [MEM 06] and the compliance of food products to safety criteria set by regulations [KOU 07].

Risk assessment is a structured process which determines the risk associated with any type of physical, chemical or biological hazard in a food and has the objective of characterizing the nature and probability of harm resulting from human exposure to agents in food. MRA provides a scientific description of foodborne risks related to the occurrence of pathogenic microorganisms in the whole food chain. Mathematical models are used to describe the introduction of pathogens into food, the growth of microorganisms in food over time, the inactivation-destruction of microorganisms by heat or other types of treatments, the consumption of microorganisms that the food is carrying and the event of the subsequent illness. MRA estimates the number of cases of (certain) illness per year, for instance 100,000 persons, in a given population caused by a certain microorganism or group of microorganisms in a particular food or food type. The output of this approach is a probability (risk), and obviously when this approach is used to answer the question "is the product safe?" the response would be a quantitative (continuous) variable. In this case, the question that arises here is "what is the degree of product safety" or "what is the acceptable risk" given that 100% of safety (zero risk) does not exist? The framework of risk analysis (steps in risk analysis: risk management, risk assessment and risk communication), which has been advocated by the World Health Organization (WTO) and Food Agricultural Organization (FAO), and the new metrics of Appropriate Level of Protection (ALOP) and Food Safety Objectives (FSO) [WTO 95] constitute a significant driving force for fair commerce and safe food trade between nations. They emerged to help the risk managers to decide and set health protection goals and use these to formulate the targets for all the relevant supply chains [GOR 05].

Given the above information, this chapter discusses how the techniques of modeling the microbial behavior can be applied in food safety management and how the predictive modeling can meet the metrics of the risk-based food safety management.

4.2. Risk-based food safety management

Recent research data shows that for developed countries that up to one quarter or even one-third of the population is affected by foodborne illness each year [GKO 13, SCA 11]. Obviously, the management of the food safety is considered a topic of fundamental public health concern. The involvement of many stakeholders in the food supply chain and the globalization of the food market have posed big challenges for national food safety authorities who are responsible for the articulation of the levels of control that the food industry has to achieve [GOR 05, FAO 05, FAO 06]. At country level, competent authorities usually apply the food safety metrics in the form of limits or criteria for microbial contaminations in food, which up till recent years have been based on food production and processing, research and expert opinions of what was considered achievable in relation to programs, systems or practices (Good Manufacturing Practices (GMPs), Good Hygiene Practices (GHPs), Good Agricultural Practices (GAPs) and HACCP) operationally available to ensure food safety [GKO 13, EFS 07]. However, the implementation of such metrics from the food production systems was difficult to be connected with the achieved public health protection. The introduction of the risk analysis by the Codex Alimentarius [COD 99] and the advances toward this field resulted in making this connection more feasible and legally required [COD 07].

For this purpose, food safety changes in legislation, such as time-temperature criteria for pasteurization and sterilization, microbiological criteria and FSOs were introduced in food safety management. All these approaches have helped us to define the specific targets or systems to improve the management of food safety [ZWI 15].

A risk-based safety management is an important step in improving food safety by linking science with safety requirements and criteria to the public health problems that they are designed to address. The metrics of the ALOP and the FSO have been proposed [FAO 04, ICM 02] to establish a link between public health outcomes and metrics in the food chain. The ALOP, introduced in the Agreement on Sanitary and Phytosanitary Measures (SPS Agreement) of the WTO is defined as: "*the level of protection deemed appropriate by the Member establishing a sanitary or phytosanitary measure to protect human, animal and plant life or health within its territory*" and represents a country's currently achieved public health status in relation to food safety and [EFS 07, WTO 95]. Nevertheless, the ALOP was not a useful measure in the actual implementation of food controls through the food chain and thus, at a later stage, ICMSF [ICM 02] introduced the FSO concept which is a measurable target for the producers, manufactures and control authorities in order to translate the ALOP into a benchmark in the food chain that could be communicated and managed by the food industry. The FSO is defined as "*the maximum frequency and/or concentration of a hazard in a food at the time of consumption that provides or contributes to the ALOP*" [COD 04].

Given that FSO is a regulatory standard which is applied at consumption time, the metrics of the performance objective (PO) and the performance criterion (PC) were created to complement the ALOP and FSO, providing targets for operational food safety management earlier in the food chain [GOR 05]. PO is the maximum frequency and/or concentration of a hazard in a food at a specified step in the food chain before the time of consumption that provides or contributes to an FSO or ALOP as applicable, while PC is the effect in frequency and/or concentration of a hazard in a food that must be achieved by the application of one or more control measures to provide or contribute to a PO or an FSO [COD 04]. The PC can be distinguished in process criteria, which are the control parameters (e.g. time-temperature conditions of a heat treatment) at a step or combination of steps that can be applied to achieve a PC, and in product criteria (PdC) which are the control parameters (e.g. salt concentration and pH) at a step or combination of steps that can be applied to achieve a PC. By definition, it is obvious that all these metrics are providing a connection between competent authorities and food business operators and therefore, they are desirable for making food safety transparent and quantifiable [ZWI 05] (Figure 4.1).

When establishing PC, consideration must be given to the initial level of a hazard and changes occurring during production, distribution, storage, preparation and use of a product. By integrating the changes in a hazard from the initial level (H_0), minus the sum of the reductions (R), plus the sum of increase (I) due to growth or recontamination, we arrive at a concentration/prevalence that at consumption time must be lower than an FSO, or a PO when it refers to a certain step, as expressed in the following equation [ICM 02]:

$$H_0 - \Sigma R + \Sigma I \leq PO \text{ (or FSO)} \qquad [4.1]$$

where H_0=Initial level of the hazard, ΣR =Total (cumulative) reduction of the hazard, ΣI =Total (cumulative) increase of the hazard, FSO, H_0, R and I are expressed in log 10 units.

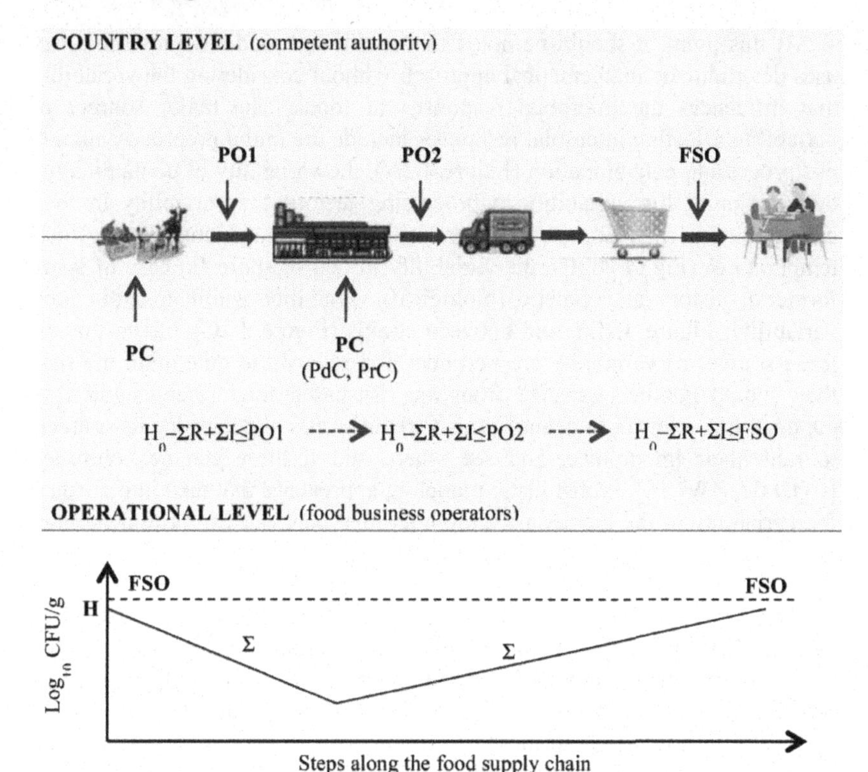

Figure 4.1. *Visual representation of the FSO, POs and PCs along the steps of the food supply chain*

4.2.1. *Predictive microbiology in a risk-based approach*

Although the framework of [4.1] is a conceptual model rather than a mathematical model (Figure 4.1), for each component of this equation the tools of predictive microbiology can be used [ZWI 05]. For example, mathematical models have been used for modeling the initial contamination level in product [BAR 05], for modeling the bacterial reduction [AND 11, CER 01, MAL 11, MEM 07, NAU 01, SOS 11], for modeling bacterial increase [GUL 06, AUG 06, KUT 05] or for modeling recontamination

[AZI 06, DEN 07, PER 11, REI 04]. Furthermore, the overall concept of [4.1] has been extensively applied in QMRA studies [AND 11, BEA 12, MEM 07, SOS 11, STE 03].

At this point, it should be noted that traditional predictive microbiology uses deterministic mathematical approach without considering the variability that influences the microbial responses in foods. The major sources of variability affecting microbial responses include the initial preprocessing and postprocessing contamination (Figure 4.2A), the variability of contamination between units, the variability in processing factors, the variability in food characteristics (pH, a_w, etc.), the variability of storage conditions (time-temperature) (Figure 4.2B), the variability in cell or spore (in case of spore formers), history and genetic (biological) variability within a strain (cell variability; Figure 4.2D), and between strains (Figure 4.2C). Given that all these sources of variability are pertinent and relevant to determine the risk, they should not be neglected from the risk-based food safety studies. In contrast, it is relevant to quantify this variability and to determine its sources, to rank their importance, and see where and if these can be controlled [COD 07, ZWI 15]. Probabilistic modeling approaches that take into account the variability of the factors affecting microbial behavior can provide a more realistic estimation of the food quality and safety.

The time-temperature conditions to which the foods are exposed at various steps during their manufacture, transportation and storage are among the most important sources of variability that are influencing the microbial responses in foods (Figures 4.2A and B). Temperature fluctuations are frequently encountered during distribution and storage of foods and are essential especially for MRA studies to be taken into account given that they affect the level of risk [KOU 10].

The inherent differences among identically treated strains of the same microbial species, also known as strain variability, constitute another important source of variability in microbiological studies [WHI 02]. This means that research data generated for a certain strain cannot be extended to other strains of the same species. For example, if the variability that characterizes the growth behavior of the strains of a foodborne pathogen, such as the one that is being presented in Figure 4.1C, is neglected in deterministic modeling approaches, such approaches may provide incomplete or even misleading predictions [POS 03]. Hence, information regarding the strain variability of phenotypic responses of foodborne pathogens under various environmental conditions is expected to be valuable when modeling microbial responses in food safety studies. Recently, in a study conducted in our laboratory [LIA 11a], aiming at characterizing the growth variability of *Salmonella enterica* as affected by the growth environment (*pH* and a_w), it was revealed that the strain variability increased as the growth conditions became more stressful, highlighting the need of incorporating this variability into MRA studies or even in studies where the factor ΣI of equation [4.1] needs to be estimated. An example of incorporation of this type of variability into predictive models of growth, transforming them into stochastic models, was the one presented by Lianou and Koutsoumanis [LIA 11b] for *Salmonella enterica.*

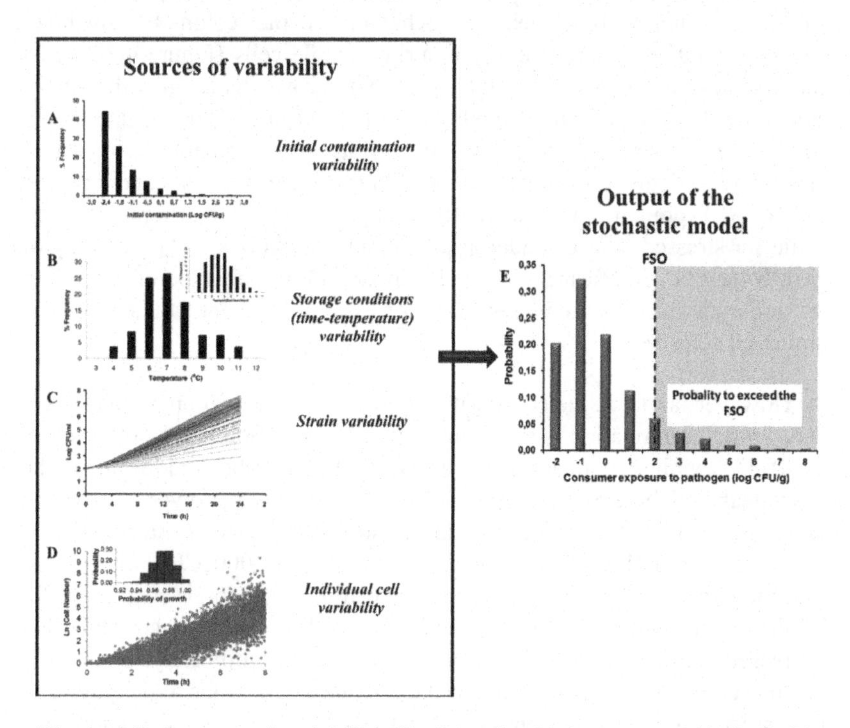

Figure 4.2. *Sources of variability affecting microbial responses (A, B, C, D) taken into account in stochastic modeling approaches in risk-based food safety studies (E). B: Adapted from [KOU 10]. C: Adapted from [LIA 11b]. D: Adapted from [KOU 13]*

Conventional bacterial growth studies use deterministic mathematical models that describe the growth of large microbial populations as a whole without considering the individual cells or without taking into account the probability of growth of the cells, assuming that all cells in the population are able to grow. However, individual cells can exhibit marked behavioral heterogeneity, and that may significantly affect the determination of the ΣI and ΣR of [4.1] and of course in the achievement of a PO or an FSO. The biological heterogeneity of individual cell division and growth has received increased attention in the recent years [GUI 06, KOU 08, KOU 13, MET 08, PIN 06, SME 08]. The division times of single cells were studied from Pin and Baranyi [PIN 06] who proposed a birth model to describe the variability in the microbial populations. Later, Koutsoumanis and Lianou [KOU 13] reported a highly heterogeneous behavior in the colonial growth of *Salmonella enterica* serotype Typhimurium single cells (Figure 4.2D). The authors developed a model that provides probabilistic growth curves indicating that the growth of single cells or small microbial population is a pool of events each one of which has its own probability to occur. Furthermore, in the last study, it was shown that under optimal conditions the cells of *S.* Typhimurium have a high probability to grow and form colonies, while for stressed cells or under less favorable conditions, the probability of growth may be significantly lower. By incorporating this type of variability in stochastic models, the credibility of risk assessment studies can be improved considerably.

However, as far as heterogeneity in microbial inactivation is concerned, very limited information is available. Recently, Aspridou and Koutsoumanis [ASP 15] applied a quantitative approach to describe and evaluate the individual cell heterogeneity (of *Salmonella enterica* serotype Agona) as variability source in microbial heat inactivation. The Monte Carlo simulation results for a population with initial concentration 10,000 CFU/ml revealed that the variability in the predicted inactivation behavior is negligible for concentrations down to 100 cells. Nevertheless, as the concentration decreased below 100 cells, the variability increased significantly. The results of this study also demonstrated that the *D*-value used so far in deterministic first-order kinetic inactivation models is valid only for large populations. On the contrary, for small populations, *D*-value presents a high variability, originating from individual cell heterogeneity, and thus, can be better characterized by a probability distribution instead of a uniform value. In addition, it was shown that the distribution of *D*-value is not the same for all population levels. The smaller the population, the more widely spread the distribution of the time required for 1 log cycle reduction is. The concept of the statistical approach proposed in the work of Aspridou and Koutsoumanis

[ASP 15] is not new, since microbial inactivation is a typical stochastic process and all the models reflect in a way the individual cell variability. Numerous studies have also considered the inactivation curve as a cumulative distribution of lethality events [COU 05, FER 99, MAF 02, PEL 98, VAN 02]. However, in all these studies, the cumulative form of individual cell inactivation time was used and applied in a deterministic way, in which the microbial population is considered as whole without taking into account and evaluating the individual cell heterogeneity as variability in the output of the models. On the basis of the generated data and information provided from the last study, it can be said that the quantification of variability can be applied in combination with deterministic models for the development of risk-based processing designs [MEM 06], especially when it is required to meet a specific PO after processing.

If the different types of variability that are affecting the microbial responses in foods are assumed to be negligible without being carefully assessed and characterized, the accuracy of risk estimates may be seriously compromised. By incorporating the various sources of variability (some of them are presented in Figure 4.2) into stochastic models, the concentration of the pathogen at consumption time can be predicted. As can be seen in Figure 4.2, the answer to the question "is the product safe?" is not a single value (yes or no) but a probability distribution that shows what is the probability of the consumer exposure to a pathogen at the time of consumption. Given that the FSO, for example, says that the pathogen will not exceed 2 logCFU/g of food when eaten, via the use of the derived distribution (Figure 4.2E) the probability of exceeding the FSO can be estimated.

Except from variability, a further complicating factor is the uncertainty. With regard to risk precision, in food safety studies, it is really important that uncertainty and variability are evaluated and expressed separately. In terms of microbial growth, uncertainty may result from the fitting of the models to the data or is associated with imprecise measurements or lack of knowledge of the effect of conditions that are not included in models, and may be reduced by additional measurements. Although approaches for separation of variability and uncertainty have been proposed [POU 03], such a task is generally difficult, and uncertainty and variability are frequently treated alike by implicitly assuming that either one or the other is negligible [NAU 00, ROS 03]. In some cases, such dissociation made by using second-order Monte Carlo simulation [DEL 06, POU 03]. However, second-order modeling requires a very high number of iterations and may become extremely time-consuming, and, thus, as an option is not attainable in the framework of a simple approach intended for the food industry.

4.2.2. *Modeling microbial responses: application to food safety*

Very often, the factor ΣI of [4.1] depends greatly on the growth of the pathogens during distribution and storage of food at retail and domestic level as well. Conditions during the above stages of the food chain are out of manufacturer's direct control and regularly deviate from specifications, and thus, they may result to increased microbial growth which revokes the efforts made for controlling the rest parameters H_0 and ΣR *via* HACCP, good manufacturing and hygiene practices. However, without scientific data, the ability of the pathogens to grow in certain food products during their shelf-life is difficult to evaluate.

On 1 January 2006, Commission Regulation (EC) 2073/2005 became effective for all European Union (EU) states [EUR 05]. Annex I of Regulation 2073 lists the microbiological criteria for foodstuffs, which are classified into food safety criteria and process hygiene criteria. Among the new food safety criteria of particular interest are those concerning *Listeria monocytogenes* in ready-to-eat (RTE) foods as, for certain food categories, they no longer require zero tolerance, but rather specify a maximum allowable concentration limit of 100 CFU/g or ml [EUR 05]. For RTE foods that are able to support the growth of *L. monocytogenes*, the Regulation demands the absence of the pathogen (in 25 g) *"before the food has left the immediate control of the food business operator, who has produced it"*, but allows for up to 100 CFU/g for *"products placed on the market during their shelf-life"*, which can be translated into an FSO (*L. monocytogenes* in RTE food product will not exceed 100 CFU/g at time of consumption). Although it seemed that these safety criteria were more lenient toward food manufactures than the previous ones, which required RTE food free of *L. monocytogenes*, it is the responsibility of the manufacturer to engage in research and generate product-specific data in order to prove scientifically that the product meets the aforementioned requirements. The regulation, by way of explanation, is driving the food industry in the direction of adapting alternative approaches to food assurance, like the use of quantitative microbiology.

The probabilistic modeling approach utilized by Koutsoumanis and Angelidis [KOU 07] for evaluating the compliance of RTE with the new European Union safety criteria constitutes a case study of how the modeling approaches can be applied in risk-based food safety management.

The study concerned RTE deli meats and during the first step of the research the ability of these RTE meat products to support the growth of *L. monocytogenes* was evaluated. The authors sampled 160 deli meat products from Greece (packaged under vacuum or modified atmosphere, stored under refrigeration) and they estimated the pH and a_w values of each product. Based on Regulation 2073/2005 for the group of products that do not support the growth of *L. monocytogenes* "*with pH* ≤ 4.4 *or* $a_w \leq 0.92$, *products with pH* ≤ 5.0 *and* $a_w \leq 0.94$" they found that only 8.2% of the tested products belonged to this category. It is obvious, that for most of the RTE meat products found in the market the food manufactures should evaluate their ability to support the growth of the pathogen. When this ability of the tested products was evaluated at 4, 10 and 15°C with the aid of a growth/no growth (G/NG) interface model, developed by Koutsoumanis and Sofos [KOU 05] for temperature, *pH* and a_w, it was predicted that the 75.6, 85 and 89.4%, respectively, of the products were able to support *L. monocytogenes* growth. The above finding indicates that although at low temperatures some products do not allow the growth of the pathogen, at higher the growth is permitted. However, from the Regulation 2073/2005, it is not clear at which temperature the food industry should evaluate this ability. The only reference to the temperature is the one presented in Article 3 where it is stated: "*Food business operators shall ensure that the food safety criteria applicable throughout the shelf life of the products can be met under reasonably foreseeable conditions of distribution, storage and use*". Thus, in the next step of the study, the temperature of 50 retail refrigerators for deli meats (in Greece) was monitored in order to calculate the *reasonably foreseeable conditions* of storage. Interestingly, it was found that the temperature varies, thus, they described it better with a distribution (Normal (4.42, 2.63)°C) for further modeling purposes.

In order to examine whether the guidelines on categorizing the products based on their ability to support or not the growth can stand in reality, the distribution of the probability of growth of *L. monocytogenes* in a given product and the percent of the product's packages in the market that are able to support growth of the pathogen were estimated. For this, a probabilistic modeling approach (Monte Carlo simulation) was used based on the following equation, in which the probability (*Pr(growth)*) of the package to support growth (1) or not (0) of the pathogen was determined.

$$Pr(growth) = \text{Binomial}(1, P_g) \quad [4.2]$$

where P_g is the probability of growth derived from the above mentioned proposed G/NG model of Koutsoumanis and Sofos [KOU 05].

The results showed that for the 160 tested products only for the 16.9% of them the percent of packages that are able to support the growth of *L. monocytogenes* is zero. However, this does not mean that the rest of them are allowing the growth of the pathogen in all the packages. For bresaola (*pH* 6.75, a_w 0.924) and for pork shoulder product (*pH* 5.49 a_w, 0.943), for example, it was found that the packages which are predicted to support *L. monocytogenes* growth were 0.1 and 33% (Figure 4.3), respectively. The latter observation for the pork shoulder emerges the question in which category this product should be listed to, in the group of RTE foods that are able to support the growth of *L. monocytogenes* or to the group that is unable to support the growth of the pathogen. Obviously, these findings are suggesting the need for guidelines on categorizing the products in a more probabilistic way.

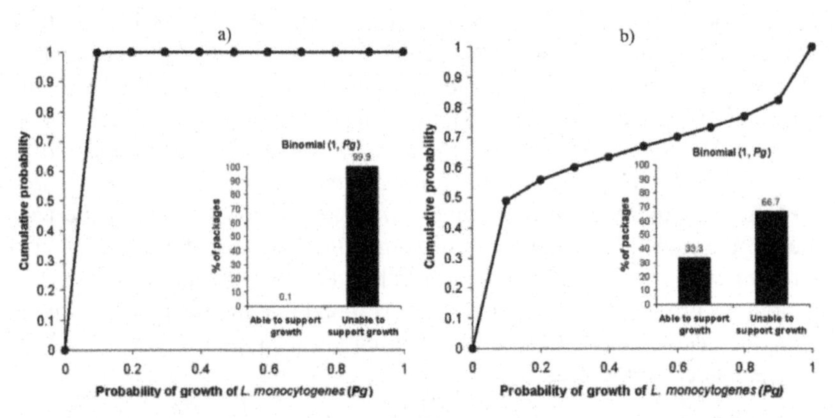

Figure 4.3. *Cumulative distribution of the probability of growth of L. monocytogenes in bresaola a) and pork shoulder b) and percent of packages that are able or unable to support growth of the pathogen during storage in retail settings (adapted from [KOU 07])*

In the second part of the study, the growth of the pathogen was evaluated until the end of the shelf-life of each product, or differently the ΣI was evaluated (Figure 4.1), and the case of meeting the FSO (2 log CFU/g at the time of consumption) was discussed.

For this purpose, [4.2] was incorporated into a kinetic three-phase linear model, the initial contamination was assumed to follow a normal distribution (Normal (9, 3.5) log(CFU/g), truncated to 2.3 log(CFU/g) based on an average package weight of 200 g) [FSI 03], the maximum concentration was assumed to be constant (10 log(CFU/g)) and with the aid of the secondary models of Buchanan and Philips [BUC 00] for the growth rate and lag phase, the distribution of the concentration of *L. monocytogenes* at the end of shelf-life was calculated using Monte Carlo simulation technique.

The probabilistic character of the modeling approach followed, by taking into account important sources of variability such as storage temperature and initial contamination, revealed the level of compliance of the shelf-life of the deli meat products with the food safety criteria of the Regulation 2073/2005. In particular, in 66.1% products tested in the research, the level of compliance was less than 50% (e.g. 66.1% of the products tested are expected to have more than 50% of their contaminated packages exceeding 100 CFU/g by the end of their shelf-life), while in only 25% of the products the level of compliance was found to be higher than 90% Table 1. Nevertheless, 100% compliance was not observed for any of the tested products. Indeed, achieving absolute (100%) compliance with the safety criterion may not be realistic, because even for contaminated products that do not support the growth of *L. monocytogenes* there is a finite probability that the initial contamination will exceed 100 CFU/g.

On the basis of the generated data, the authors presented two concepts in order to achieve a certain level of compliance. The first one was relied on the modification of the shelf-life of the products. As can be seen in Figure 4.4, if for the aforementioned pork shoulder product the shelf-life is going to be decreased from 113 to 50 or 36 days, the level of compliance with the safety criteria is going to be increased from 64.7 to 90% and 95%, respectively.

The second concept was based on the proposal for modification of the formulation of the product, or differently, for modification of the PdC. A decreasing of the a_w of the product from 0.943 to 0.930 and increasing the concentration of $NaNO_2$ from 50 to100 ppm would result in 90% compliance for the same shelf-life of 133 days (Figure 4.5). Obviously, the above approach can be utilized by the food industry and constitutes a basis for a risk-based management.

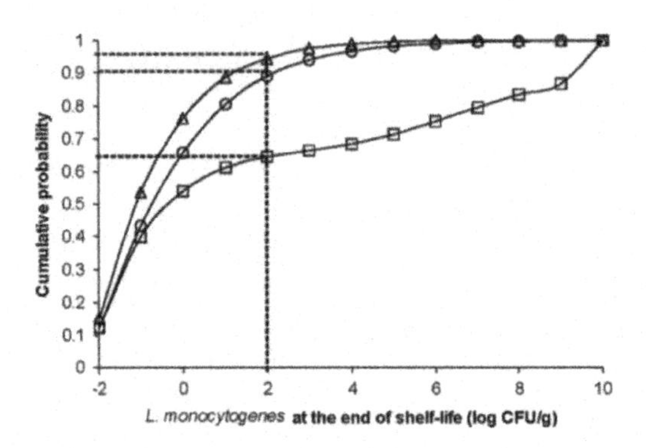

Figure 4.4. *Effect of shelf-life modifications on the cumulative probability distribution of the L. monocytogenes concentration in contaminated pork shoulder packages at the end of shelf life. □, current shelf-life of 113 days; ○, shelf-life of 50 days; Δ, shelf-life of 36 days. Dotted lines indicate the level corresponding to compliance with the 100 CFU/g safety criterion (adapted from [KOU 07])*

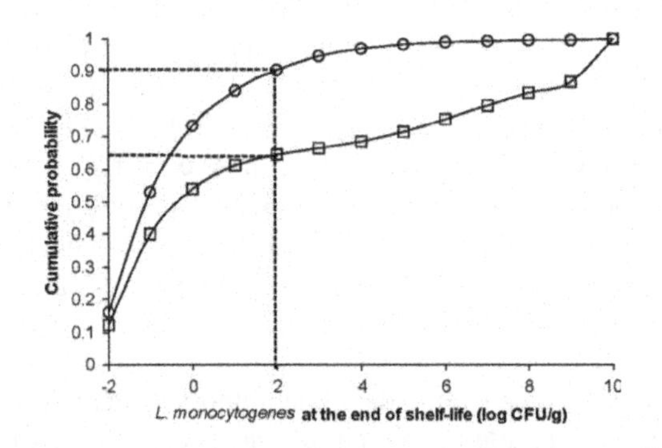

Figure 4.5. *Effect of modifications of product formulation on the cumulative probability distribution of L. monocytogenes concentration in contaminated pork shoulder packages at the end of the shelf-life.□, current formulation (pH 5.49, a_w 0.943, $NaNO_2$ 50 ppm); ○, modified formulation (pH 5.49, a_w 0.930, $NaNO_2$ 100 ppm). Dotted lines indicate the level corresponding to compliance with the new safety criteria (adapted from [KOU 07])*

4.3. Conclusion

Available data and research studies in predictive microbiology have allowed the use of the models in the food safety management. Predictive models have been used in order to meet the metrics of the risk-based safety management. However, despite the fact that nowadays the use of ALOP and

FSO in managing food safety is clear, so far no country applies the targets as such to inform national policy although some countries have established other specific targets in the food chain aiming at the reduction of their current disease burden [GKO 13]. A reason for this can be that there is limited assistance on how to establish these targets [HAV 04] and how to implement them in practice linked to each other [RIE 07, STR 05] although several documents have been published so far providing guidance and ways of implementation on the use of the metrics for specific pathogen/food combinations [BUC 06, BUT 06, DEP 06, FAO 06, GKO 13, POU 09, TRO 10, TUO 07, AND 11, CRO 09, MEM 07, RIE 07, SOS 11, STE 03]. Therefore, it is considered essential to invest in the future on the strategies that connect risk assessors with the risk managers.

4.4. Bibliography

[AND 89] ANDERSON N.M., LARKIN J.W., COLE M.B. *et al.*, "Food safety objective approach for controlling Clostridium botulinum growth and toxin production in commercially sterile foods", *Journal of Food Protection*, vol. 74, pp. 1956–1989, 1989.

[ARM 97] ARMITAGE N.H., "Use of predictive microbiology in meat hygiene regulatory activity", *International Journal of Food Microbiology*, vol. 36, pp. 103–109, 1997.

[ASP 15] ASPRIDOU Z., KOUTSOUMANIS K.P., "Individual cell heterogeneity as variability source in population dynamics of microbial inactivation", *Food Microbiology*, vol. 45, pp. 216–221, 2015.

[AZI 06] AZIZA F., METTLER E., DAUDIN J.-J. *et al.*, "Stochastic, compartmental, and dynamic modeling of cross-contamination during mechanical smearing of cheeses", *Risk Analysis*, vol. 26, pp. 731–745, 2006.

[BAR 05] BARKER G.C., MALAKAR P.K., DEL TORRE M. *et al.*, "Probabilistic representation of the exposure of consumers to Clostridium botulinum neurotoxin in a minimally processed potato product", *International Journal of Food Microbiology*, vol. 100, pp. 345–357, 2005.

[BEA 12] BEAN D., BOURDICHON F., BRESNAHAN D. *et al.*, Risk assessment approaches to setting thermal processes in food manufacture, ILSI Europe Report Series, ILSI Europe Risk Analysis in Food Microbiology Task Force, 2012.

[BUC 00] BUCHANAN R.L., PHILLIPS J.G., "Updated models for the effects of temperature, initial pH, NaCl and NaNO2 on the aerobic and anaerobic growth of *Listeria monocytogenes*", *Quant. Microbiol.*, vol. 2, pp. 103–128, 2000.

[BUC 06] BUCHANAN R., WHITING R., ROSS T., Background paper for the joint FAO/WHO expert consultation on development of practical risk management strategies based on microbiological risk assessment outputs, Kiel, Germany, 3–7 April 2006 [Case study: *Listeria monocytogenes* in smoked fish. "Development of Risk Management Metrics for Food Safety". Food and Agriculture Organization of the United Nations/World Health Organization, Rome/Geneva].

[BUT 06] BUTLER F., DUFFY G., ENGELJOHN D. *et al.*, Background paper for the joint FAO/WHO expert consultation on the development of practical risk management strategies based on microbiological risk assessment outputs, Kiel, Germany, 3–7 April 2006. [Case study: *Escherichia coli* O157:H7 in fresh ground beef. Food and Agriculture Organization of the United Nations/World Health Organization, Rome/Geneva].

[CAS 98] CASSIN M.H., LAMMERDING A.M., TODD E.C.D. *et al.*, "Quantitative risk assessment for *Escherichia coli* O157:H7 in ground beef hamburgers", *Int. J. Food Microbiol.*, vol. 41, pp. 21–44, 1998.

[CER 01] CERF O., DAVEY K.R., "An explanation of non-sterile (leaky) milk packs in well- operated UHT plant", *Food and Bioproducts Processing*, vol. 79, pp. 219–222, 2001.

[COD 99] CODEX ALIMENTARIUS, Principles and guidelines for the conduct of microbiological risk assessment, CAC/GL-30 Codex Alimentarius Commission, Rome, 1999.

[COD 04] CODEX ALIMENTARIUS, "Codex Alimentarius Commission", in ALIMENTARIUS C. (ed.), *Procedural Manual*, 14th ed., World Health Organization/Food and Agriculture Organization of the United Nations, Geneva/Rome, 2004.

[COD 07] CODEX ALIMENTARIUS, Working principles for risk analysis for food safety for application by governments, CAC/GL 62–2007, World Health Organization/Food and Agriculture Organization of the United Nations, Geneva/Rome, 2007.

[COU 05] COUVERT O., GAILLARD S., SAVY N. *et al.*, "Survival curves of heated bacterial spores: effect of environmental factors on Weibull parameters", *Int. J. Food Microbiol.*, vol. 101, pp. 73–81, 2005.

[CRO 09] CROUCH E.A., LABARRE D., GOLDEN N.J. *et al.*, "Application of quantitative microbial risk assessments for estimation of risk management metrics: *Clostridium perfringens* in ready-to-eat and partially cooked meat and poultry products as an example", *Journal of Food Protection*, vol. 72, pp. 2151–2161, 2009.

[DEL 06] DELIGNETTE-MULLER M.L., CORNU M., POUILLOT R. *et al.*, "Use of Bayesian modelling in risk assessment: application to growth of *Listeria monocytogenes* and food flora in cold-smoked salmon", *International Journal of Food Microbiology*, vol. 106, pp. 195–208, 2006.

[DEN 03] DEN AANTREKKER E.D., BOOM R.M., ZWIETERING M.H. *et al.*, "Quantifying recontamination through factory environments – a review", *International Journal of Food Microbiology*, vol. 80, pp. 117–130, 2003.

[DEP 06] DEPAOLA A., LEE R., MAHONEY D. *et al.*, Background paper for the joint FAO/WHO expert consultation on the development of practical risk management strategies based on microbiological risk assessment outputs, Kiel, Germany, 3–7 April 2006 [Case study: Vibrio vulnificus in oysters. Food and Agriculture Organization of the United Nations/World Health Organization, Rome/Geneva].

[EFS 07] EFSA/ECDC, "The community summary report on trends and sources of zoonoses, zoonotic agents, antimicrobial resistance and foodborne outbreaks in the European Union in 2006", *EFSA Journal 5*, 2007.

[EUR 05] EUROPEAN COMMISSION, "Commission Regulation (EC) No 2073/2005 of 15 November 2005 on microbiological criteria for foodstuffs", *Official Journal of the European Union*, vol. L 338, pp. 1–26, 2005.

[FAO 04] FAO/WHO (FOOD AND AGRICULTURE ORGANIZATION OF THE UNITED NATIONS/WORLD HEALTH ORGANIZATION), "Risk assessment of *Listeria monocytogenes* in ready-to-eat foods" *MRA*, series 4–5, WHO, available at: http://www.who.int/foodsafety/publications/micro/mra_listeria/en/print.html. Accessed 26.05.14, 2004.

[FAO 06] FAO/WHO, Food safety risk analysis. A guide for national food safety authorities, FAO Food and Nutrition Papers, Food and Agriculture Organization of the United Nations/World Health Organization, Rome/Geneva, 2006.

[FDA 03] FDA/FSIS JOINT EXPERT CONSULTATION, Quantitative assessment of the relative risk to public health from foodborne *Listeria monocytogenes* among selected categories of ready-to-eat foods, U.S. Department of Agriculture Food Safety and Inspection Service, Washington, DC, available at: http://www.foodsafety.gov/ dms/lmr2-toc.html, 2003.

[FER 99] FERNANDEZ A., SALMERON C., FERNANDEZ P.S. *et al.*, "Application of a frequency distribution model to describe the thermal inactivation of two strains of *Bacillus cereus*", *Trends Food Sci. Technol.* vol. 10, pp. 158–162, 1999.

[FSI 03] FSIS, FSIS risk assessment for *Listeria monocytogenes* in deli meats, Food Safety and Inspection Service, U.S. Department of Agriculture, Washington, DC, 2003.

[GIA 01] GIANNAKOUROU M., KOUTSOUMANIS K., NYCHAS G.-J. E. *et al.*, "Development and assessment of an intelligent shelf life decision system (SLDS) for chill chain optimization", *J. Food Prot.*, vol. 64, pp. 1051–1057, 2001.

[GKO 13] GKOGKA E., REIJ M.W., GORRIS L.G.M. *et al.*, "Risk assessment strategies as a tool in the application of the Appropriate Level of Protection (ALOP) and Food Safety Objective (FSO) by risk managers", *International Journal of Food Microbiology*, vol. 167, no. 2013 pp. 8–28, 2013.

[GOR 05] GORRIS L.G.M., "Food safety objective: an integral part of food chain management", *Food Control*, vol. 16, pp. 801–809, 2005.

[GUI 06] GUILLIER L., AUGUSTIN J.C., "Modelling the individual cell lag time distributions of Listeria monocytogenes as a function of the physiological state and the growth conditions", *International Journal of Food Microbiology*, vol. 111, pp. 41–251, 2006.

[HAV 04] HAVELAAR A.H., NAUTA M.J., JANSEN J.T., "Fine-tuning Food Safety Objectives and risk assessment", *International Journal of Food Microbiology*, vol. 93, pp. 11–29, 2004.

[ICM 02] ICMSF (INTERNATIONAL COMMISSION ON MICROBIOLOGICAL SPECIFICATIONS FOR FOODS), *Microorganisms in Foods, Book 7, Microbiological Testing in Food Safety Management*, Kluwer Academic/Plenum, NY, 2002.

[JUN 03] JUNEJA V.K., MARKS H.M., HUANG L., "Growth and heat resistance kinetic variation among various isolates of Salmonella and its application to risk assessment", *Risk Anal.*, vol. 23, pp. 199–213, 2003.

[KOS 11] KOSEKI S., MIZUNO Y., SOTOME I., "Modeling of pathogen survival during simulated gastric digestion", *Appl. Environ. Microbiol.*, vol. 77, pp. 1021–1032, 2011.

[KOU 02] KOUTSOUMANIS K., GIANNAKOUROU P.S. TAOUKIS *et al.*, "Application of shelf life decision system (SLDS) to marine cultured fish quality", *Int. J. Food Microbiol.*, vol. 73, pp. 375–382, 2002.

[KOU 05a] KOUTSOUMANIS K., SOFOS J.N., "Effect of inoculum size on the combined temperature, pH and aw limits for growth of *Listeria monocytogenes*", *Int. J. Food Microbiol.*, vol. 104, pp. 83–91, 2005.

[KOU 05b] KOUTSOUMANIS K., TAOUKIS P.S., NYCHAS G.-J.E., "Development of a safety monitoring and assurance system (SMAS) for chilled food products", *Int. J. Food Microbiol.*, vol. 100, pp. 253–260, 2005.

[KOU 07] KOUTSOUMANIS K., ANGELIDIS A.S., "Probabilistic modeling approach for evaluating the compliance of ready -to- eat foods with new European Union safety criteria for *Listeria monocytogenes*", *Appl. Environ. Microbiol.*, vol. 73, pp. 4996–5004, 2007.

[KOU 08] KOUTSOUMANIS K., "A study on the variability in the growth limits of individual cells and its effect on the behaviour of microbial populations", *Int. J. Food Microbiol.*, vol. 128, pp. 116–121, 2008.

[KOU 10] KOUTSOUMANIS K., PAVLIS A., NYCHAS G.-J.E. *et al.*, "Probabilistic model for *Listeria monocytogenes* growth during distribution, retail storage, and domestic storage of pasteurized milk", *Applied and Environmental Microbiology*, vol. 76, pp. e2181–e2191, 2010.

[KOU 13] KOUTSOUMANIS K., LIANOU A., "Stochasticity in colonial growth dynamics of individual bacterial cells", *Appl. Environ. Microbiol.*, vol. 79, pp. 2294–2301, 2013.

[KOU 15] KOUTSOUMANIS K., GOUGOULI M., "Use of Time Temperature Integrators in food safety management", *Trends in Food Science and Technology*, vol. 43, pp. 236–244, 2015.

[KUT 05] KUTALIK Z., RAZAZ M., ELFWING A. *et al.*, "Stochastic modeling of individual cell growth using flow chamber microscopy images", *International Journal of Food Microbiology*, vol. 105, pp. 177–190, 2005.

[LIA 11a] LIANOU A., KOUTSOUMANIS K., "Effect of the growth environment on the strain variability of *Salmonellaenterica* kinetic behavior", *Food Microbiol.*, vol. 28, pp. 828–837, 2011.

[LIA 11b] LIANOU A., KOUTSOUMANIS K., "A stochastic approach for integrating strain variability in modeling *Salmonella enterica* growth as a function of pH and water activity", *Int. J. Food Microbiol.*, vol. 149, pp. 254–261, 2011.

[MAF 02] MAFART P., COUVERT O., GAILLARD S. *et al.*, "On calculating sterility in thermal preservation methods: application of Weibull frequency distribution model", *Int. J. Food Microbiol.*, vol. 72, pp. 107–113, 2002.

[MAL 11] MALAKAR P.K., BARKER G.C., PECK M.W., "Quantitative risk assessment for hazards that arise from non-proteolytic *Clostridium botulinum* in minimally processed chilled dairy-based foods", *Food Microbiology*, vol. 28, pp. 321–330, 2011.

[MCM 97] MCMEEKIN T.A., BROWN J.L., KRIST K. *et al.*, "Quantitative microbiology: a basis for food safety", *Emerging Infect. Dis.*, vol. 3, pp. 541–550, 1997.

[MEM 06] MEMBRÉ J.M., ZQUITA A.A., BASSETT J. *et al.*, "A probabilistic modelling approach in thermal inactivation: estimation of post-process *Bacillus cereus* spore prevalence and concentration", *Journal of Food Protection*, vol. 69, pp. 118–129, 2006.

[MEM 07] MEMBRÉ J.M., BASSETT J., GORRIS L.G.M., "Applying the Food Safety Objective and related standards to thermal inactivation of *Salmonella* in poultry meat", *Journal of Food Protection*, vol. 70, pp. 2036–2044, 2007.

[MET 08] METRIS A., GEORGE S.M., MACKEY B.M. *et al.*, "Modelling the variability of the lag times of single cells of *Listeria innocua* populations in response to sublethal and lethal heat treatments", *Appl. Environ. Microbiol.*, vol. 74, pp. 6949–6955, 2008.

[NAU 00] NAUTA M.J., "Separation of uncertainty and variability in quantitative microbial risk assessment models", *International Journal of Food Microbiology*, vol. 57, pp. 9–18, 2000.

[NAU 01] NAUTA M.J., A modular process risk model structure for quantitative microbiological risk assessment and its application in an exposure assessment of *Bacillus cereus* in a REPFED.RIVM report 149106 007, 2001.

[NIC 96] NICOLAÏ B.M., VAN IMPE J.F., "Predictive food microbiology: a probabilistic approach", *Math. Comput. Simul.*, vol. 42, pp. 287–292, 1996.

[PEL 98] PELEG M., COLE M.B., "Reinterpretation of microbial survival curves", *Crit. Rev. Food Sci. Nutr.*, vol. 38, p. 353, 1998.

[PER 11] PEREZ RODRIGUEZ F., CAMPOS D., RYSER E.T. *et al.*, "A mathematical risk model for *Escherichia coli* O157:H7 cross-contamination of lettuce during processing", *Food Microbiology*, vol. 28, pp. 694–701, 2011.

[PIN 06] PIN C., BARANYI J., "Kinetics of single cells: observation and modelling of a stochastic process", *Appl. Environ. Microbiol.*, vol. 72, pp. 2163–2169, 2006.

[POS 03] POSCHET F., GEERAERD A.H., SCHEERLINCK N. *et al.*, "Monte Carlo analysis as a tool to incorporate variation on experimental data in predictive microbiology", *Food Microbiol.*, vol. 20, pp. 285–295, 2003.

[POU 03] POUILLOT R., ALBERT I., CORNU M. *et al.*, "Estimation of uncertainty and variability in bacterial growth using Bayesian inference. Application to *Listeria monocytogenes*", *International Journal of Food Microbiology*, vol. 81, pp. 87–104, 2003.

[POU 09] POUILLOT R., GOULET V., DELIGNETTE-MULLER M.L. *et al.*, "Quantitative risk assessment of *Listeria monocytogenes* in French cold-smoked salmon: II. Risk characterization", *Risk Analysis*, vol. 29, pp. 806–819, 2009.

[REI 04] REIJ M.W., DEN AANTREKKER E.D., "ILSI Europe Risk Analysis in Microbiology Task Force, 2004. Recontamination as a source of pathogens in processed foods", *International Journal of Food Microbiology*, vol. 91, pp. 1–11, 2004.

[RIE 07] RIEU E., DUHEM K., VINDEL E. *et al.*, "Food Safety Objectives should integrate the variability of the concentration of pathogen", *Risk Analysis,* vol. 27, pp. 373–386, 2007.

[ROS 03] ROSS T., MCMEEKIN T.A., "Modeling microbial growth within food safety risk assessments", *Risk Analysis,* vol. 23, pp. 179–197, 2003.

[SCA 11] SCALLAN E., HOEKSTRA R.M., ANGULO F.J. *et al.*, "Foodborne illness acquired in the United States — major pathogens", *Emerging Infectious Diseases,* vol. 17, pp. 7–22, 2011.

[SME 08] SMELT J.P.P.M., BOS A.P., KORT R. *et al.*, "Modelling the effect of sub(lethal) heat treatment of Bacillus subtilis spores on germination rate and outgrowth to exponentially growing vegetative cells", *Int. J. Food Microbiol.*, vol. 128, pp. 34–40, 2008.

[SOS 11] SOSA MEJIA Z., BEUMER R.R., ZWIETERING M.H., "Risk evaluation and management to reaching a suggested FSO in a steam meal", *Food Microbiology*, vol. 28, pp. 631–638, 2011.

[STE 03] STEWART C.M., COLE M.B., SCHAFFNER D.W., "Managing the risk of Staphylococcal food poisoning from cream-filled baked goods to meet a Food Safety Objective", *Journal of Food Protection,* vol. 66, pp. 1310–1325, 2003.

[STR 05] STRINGER M., "Food safety objectives — role in microbiological food safety management", *Food Control,* vol. 16, pp. 775–794, 2005.

[TRO 10] TROMP S.O., FRANZ E., RIJGERSBERG H. *et al.*, "A model for setting performance objectives for *Salmonella* in the broiler supply chain", *Risk Analysis*, vol. 30, pp. 94–951, 2010.

[TUO 07] TUOMINEN P., RANTA J., MAIJALA R., "Studying the effects of POs and MCs on the *Salmonella* ALOP with a quantitative risk assessment model for beef production", *International Journal of Food Microbiology*, vol. 118, pp. 35–51, 2007.

[VAN 02] VAN BOEKEL M.A.J.S., "On the use of the Weibull model to describe thermal inactivation of microbial vegetative cells", *Int. J. Food Microbiol.*, vol. 74, pp. 139–159, 2002.

[WHI 02] WHITING R.C., GOLDEN M.H., "Variation among *Escherichia coli* O157:H7 strains relative to their growth, survival, thermal inactivation, and toxin production in broth", *International Journal of Food Microbiology*, vol. 75, pp. 127–133, 2002.

[WTO 95] WORLD TRADE ORGANIZATION, Agreement on the application of sanitary and phytosanitary measures (SPS Agreement), 1995.

[ZWI 05] ZWIETERING M., "Practical considerations on food safety objectives", *Food Control,* vol. 16, pp. 817–823, 2005.

[ZWI 15] ZWIETERING M.H., "Risk assessment and risk management for safe foods: assessment needs inclusion of variability and uncertainty, management needs discrete decisions", *International Journal of Food Microbiology*, available at: http://dx.doi.org/10.1016/j.ijfoodmicro.2015.03.032, 2015.

Conclusion

Modeling in food microbiology encompasses not only mathematical tools but also probabilistic ones.

The mathematical tools used to describe the effect of factors associated with food process, formulation (e.g. heat-treatment and pH) and distribution (e.g. storage temperature) have been described in Chapter 2. The latest predictive model developments and current available software have been listed and discussed in the context of their further use in food spoilage and food safety.

The probabilistic tools used to describe the propagation of microbial contamination (prevalence and concentration) from the farm to the fork have been described in Chapter 3. The main probabilistic distributions (Binomial, Beta, Poisson, etc.) utilized in exposure and risk assessment to propagate the uncertainty and variability inherent to biological processes have been introduced and explained through basic, but food safety-related, examples. Next, Monte Carlo simulations, uncertainty and variability have been presented.

In Chapter 4, application of modeling for solving complex food spoilage issue has been illustrated using two crucial applications: commercial sterility on the one hand and best-before-date determination on the other hand. Last, in Chapter 5, application of modeling to food safety has been presented through applications of risk-based safety management (FSO, PO and PC). In both chapters, food contamination propagation has been described using

---

Conclusion written by Jeanne-Marie MEMBRÉ and Vasilis VALDRAMIDIS.

mathematical models and probability processes, tools which are essential when trying to make predictions and/or decisions on food safety and quality.

The aim of this book was to benefit scientists and engineers who have to quantify the microbial propagation in food transformation and storage while taking into account issues that are not widely addressed yet, like the separation of uncertainty and variability. All the examples chosen to illustrate the modeling aspects are representative, and realistic, of microbial spoilage and safety associated with food transformation and distribution. They can be further used and applied by industry and stakeholders. This book will then contribute to moving toward quantitative and scientific-based food safety management in the industry.

List of Authors

Enda CUMMINS
University College Dublin
Agriculture and
Food Science Centre
Dublin
Ireland

Stéphane DAGNAS
UMR-INRA 1014 Secalim–Oniris
Nantes
France

Maria GOUGOULI
University of Thessaloniki
Department of Food Science
and Technology
Thessaloniki
Greece

Kostas KOUTSOUMANIS
University of Thessaloniki
Department of Food Science
and Technology
Thessaloniki
Greece

Jeanne-Marie MEMBRÉ
UMR-INRA 1014 Secalim–Oniris
Nantes
France

Vasilis VALDRAMIDIS
University of Malta
Msida
Malta

Index

Printed in the United States
By Bookmasters